589.2
PU
SM
KITCHENER PUBLIC LIBRARY
3 9098 01557726 4
ooms

635.8 Purse
Pursey, Helen L
The wonderful world of
mushrooms and other fungi
Kitchener Public Library
Country Hills-Nonfic

ST. MARY'S HIGH SCHOOL

ST. MARY'S SENIOR GIRLS' SCHOOL
35 WEBER STREET WEST
KITCHENER

7.95

The Wonderful World of MUSHROOMS and other FUNGI

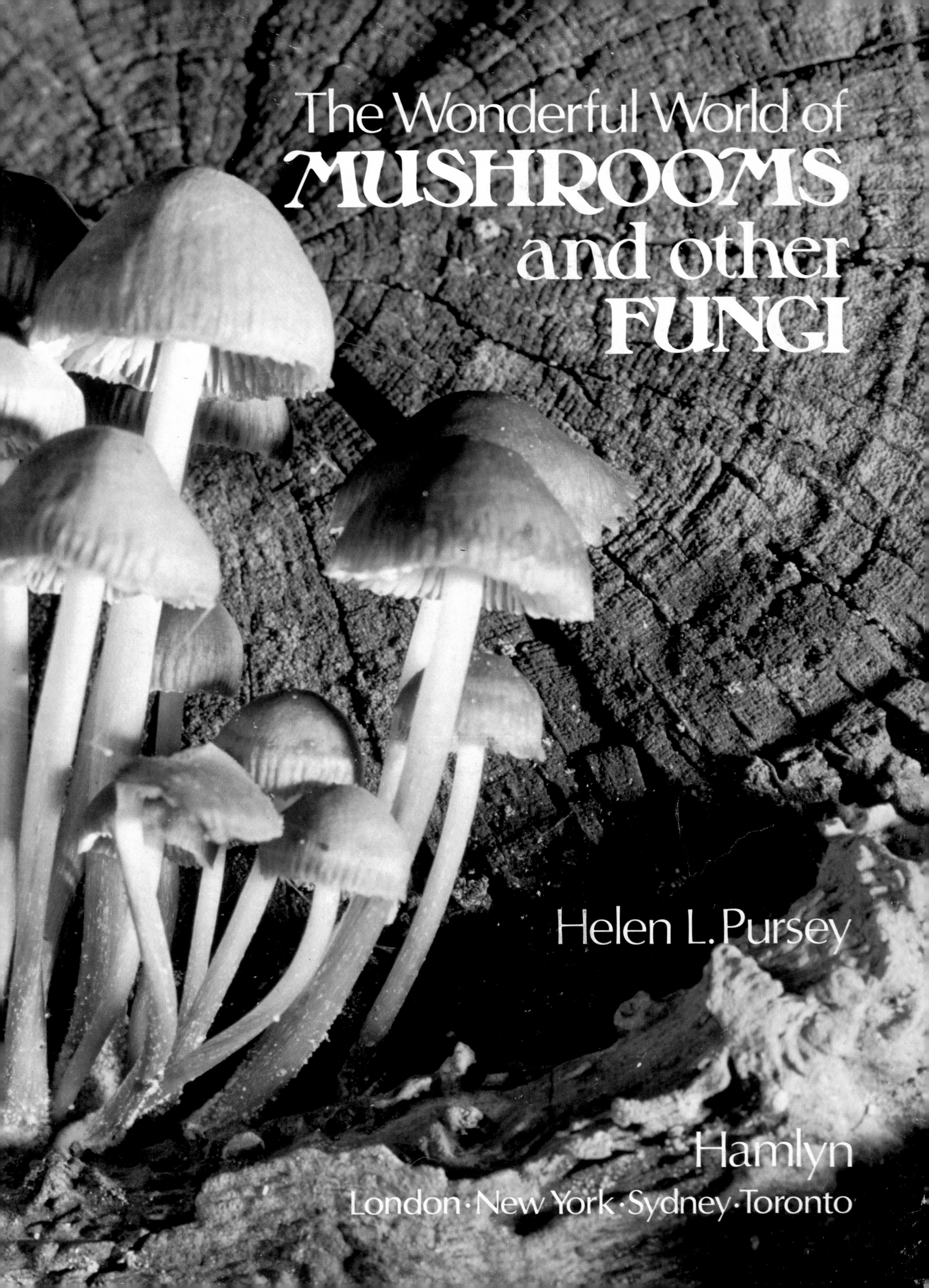

The Wonderful World of MUSHROOMS and other FUNGI

Helen L. Pursey

Hamlyn
London · New York · Sydney · Toronto

Acknowledgments

Biofotos: Heather Angel 9, 10, 11, 16, 17, 22, 23, 24T, 25, 29, 30, 32T, 35L, 36B, 37B, 38, 40, 44BR, 49, 51, 54, 59B, 64, 70, 77, 79, 81, 86B, 90, 93T; Gordon Dickson 55B, 67, 68, 87. **Bruce Coleman Limited:** John Markham 76L, 86T; R. K. Murton 18; Charles J. Ott 14C; S. C. Porter 31B, 45, 60B, 71T, 72, 75T, 82, 83. **W. F. Davidson:** 13, 19, 20, 21, 24B, 26, 28B, 34, 41, 48, 55T, 69B, 73, 84B, 94B. **Jacana:** 37T. **Ministry of Agriculture, Fisheries & Food:** 94T. **Mushroom Growers Association:** 63. **Natural History Photographic Agency:** H. R. Allen 14T, 33, 42T; Stephen Dalton 32B; Brian Hawkes *back jacket flap*, 7, 12, 15, 36T, 53, 57, 59T, 59C, 60C, 62, 65, 66, 69T, 71B, 78, 84T, 85, 91, 92, 93B; A. Huxley 27, 41; G. E. Hyde 35R, 43, 47T, 50, 52, 58, 74, 88; K. G. Preston-Mafham *front jacket, back jacket*, 8, 14B, 28T, 39, 42B, 44T, 44BL, 56, 61, 76R; Ivan Polunin *endpapers, title page*, 89T; M. I. Walker 9TR, 89B, 95. **Natural Science Photos:** David Ward *front jacket flap*, 60T, 75.

The recipe for Green Collared Trout which appears on p. 62 is reproduced with the kind permission of the Mushroom Growers Association

The photograph of Potato Blight which appears on p. 94 is Crown Copyright, and is reproduced by permission of the Controller of Her Majesty's Stationery Office

Filmset in England by Keyspools Limited,
Golborne, Lancashire
Set in 12 pt Optima

Published by the Hamlyn Publishing Group Limited
London · New York · Sydney · Toronto
Astronaut House, Feltham, Middlesex, England
Copyright © The Hamlyn Publishing Group Limited 1977

ISBN 0 600 36248 5

Printed in Spain by Mateu Cromo, S. A. Pinto (Madrid)

14037

Contents

Introduction

Of all plants probably no group is more fascinating than the fungi. They form an enormous assemblage of very varied species, differing markedly in appearance, and growing in a variety of habitats. Yet they have one all-important feature in common—they are, without exception, totally lacking in the green pigment chlorophyll which is so characteristic of, and essential to, practically all other plants.

Green plants (i.e. those with chlorophyll) have small bodies known as chloroplasts in the cells of their leaves. These chloroplasts contain the pigment chlorophyll which enables the plants to manufacture sugars and starch in the presence of light by the process known as photosynthesis. It is one of the marvels of nature that from the simple inorganic compounds carbon dioxide and water, such complex organic compounds as carbohydrates can be built up. Photosynthesis takes place in all green cells which are exposed to light, and all other organisms are ultimately dependent on this process. Green plants provide, directly or indirectly, food for man, his livestock and all other creatures. Fungi resemble animals in that they too are dependent on green plants for food.

Some fungi attack other plants (less often animals), living on their tissues and growing at their expense, usually causing disease or deformity in the process. These are parasites, and many create serious problems for man. Other fungi live on dead plant or animal remains, breaking down and absorbing the proteins and other complex substances present and helping to bring about decay. This is a natural and very necessary process since without the joint action of fungi, bacteria and a variety of small animals such as worms, beetles and flies the earth would become littered with dead plants and animals. Furthermore by bringing about this breakdown of organic matter, essential elements are recycled and made available for new generations of plants and animals.

Some fungi enter into mutually beneficial partnerships with other organisms, usually green plants. Such an association is known as symbiosis. The best-known of these associations is that of the lichens, which can be regarded as fungus-alga partnerships. Somewhat less obvious is the relationship between forest trees and a number of the larger fungi, and that between orchids and certain fungi. Both of these associations can be detected by examining the roots which, if cut in cross-section and viewed down a microscope, will be seen to have the outer layers infected with fungal hyphae. Such fungus-roots are termed mycorrhizas and, as we shall see later, play a very necessary part in the life of the plant. We now know that mycorrhizal fungi secrete natural antibiotics in the soil and these can destroy other soil fungi which attack and damage plant roots. Fungi and bacteria form many partnerships in the soil; the bacteria help to break down organic matter, making it available for the fungi, and these in turn provide essential growth substances which the bacteria lack. Some fungi enter the roots of trees and shrubs such

The bluish or yellowish-green tints of the cap and stalk of the Verdigris Agaric (*Stropharia aeruginosa*) are unusual. The colour (which is not associated with chlorophyll) is in a slimy coating which gradually washes away.

as alder and sea buckthorn, causing large clustered swellings or nodules to form. Certain bacteria induce somewhat similar nodules on the roots of leguminous plants such as beans and lupins. Within the nodules both the bacteria and fungi can transform nitrogen into complex compounds such as amino-acids which are utilized by the root cells.

The origin of the fungi is obscure. They may possibly have evolved from simple algae, and certainly this development took place in water. It is obvious that they are an extremely ancient group, since traces of them are known from Pre-Cambrian rocks which are at least 600 million years old, and probably considerably older. Fungi have a very delicate structure and do not fossilize particularly well, so their preservation in such very old rocks is remarkable. In more recent rocks obvious parasitic fungi have been found in the cells of higher plants. Recognizable fungi of all the major groups are known from rocks about 300 million years old; they resemble many of the genera found today and show several well-preserved structures. In coal measure plants, a number of which were sizeable forest trees, mycorrhizas were clearly established.

The actual number of species of fungi is a matter of some dispute. If we exclude fungus-like groups which for various reasons are not true fungi, then estimates range from 50,000 to nearly 100,000 species. There is no other group about which there is so much uncertainty as to its actual size and the only larger group is the flowering plants with about 285,000 species.

With a few exceptions fungi have a vegetative body called the mycelium. This usually takes the form of a delicate weft of threads or hyphae and on it develop the reproductive structures or fruitbodies. In the higher fungi these bodies are relatively large and conspicuous and in some species form the familiar mushrooms and toadstools, for instance, but many of the simpler fungi are extremely small, even microscopic, and thus may not be at all obvious. Quite a number of soil fungi (also aquatic fungi) living in both fresh and sea water, are invisible to the naked eye. Aquatic fungi actually form less than 2 per cent of total fungal species, the rest being terrestrial. However, fungal spores together with pollen grains, both of which are minute and dust-like, are extremely abundant in the air and constitute what is often called the air spora. The mycelium develops when a spore settles on a suitable medium for growth and begins to germinate. Moisture is essential for this process and delicate hyphae quickly begin to radiate out from the spore, forming a roughly circular mycelium. The hyphae branch repeatedly, spreading over the substrate and secreting enzymes which digest the organic matter. The soluble products are then absorbed. Soon, reproductive structures begin to form, and these give rise to new spores.

In a few fungi the hyphae aggregate to form tough cord-like structures known as rhizomorphs. These can travel for several metres through soil, rubble, brickwork and even drains, thus spreading the fungus. Two well-known examples are the Bootlace Fungus or Honey Agaric (*Armillaria mellea*) which attacks and kills standing trees as well as living on fence posts and other timber, and *Serpula (Merulius) lacrymans* which causes dry rot. More commonly hyphae may compact together to form a hard resting body known as a sclerotium. These vary in size from a millimetre or so across to large tropical types which may reach the size of a man's head.

Fungi produce a great variety of spores and often in considerable numbers. They may be produced sexually, or without the intervention of sexual processes. Asexual spores are particularly common in the lower fungi, especially the water moulds. Such spores may be found within the stalked aerial globular sporangia of the common Pin Mould (*Mucor*), or in submerged elongate sporangia as in *Saprolegnia*, in which the released spores are mobile. Asexual spores may also be budded off in chains from a parent structure of varying shape, depending on the species. Such spores are termed conidia.

Sexual processes are extremely variable and the resultant spores are used to distinguish major groups of fungi. Thus the mixed, but rather simple group Phycomycetes may be further subgrouped. One

Although common, the twiggy bright yellow Fairy Club (*Clavaria corniculata*) can easily be overlooked since it grows less than 6 cm (about $2\frac{1}{2}$ in.) high, usually hidden in grass. Ascospores are released from most of the surface except the base.

group have a globular female sex organ which is fertilized by motile male gametes from an elongate male sex organ, whereas the second group form far less well differentiated sex organs. These usually take the form of pairs of swollen side-branches which join, their contents then acting as gametes and fusing. Thus both groups have an obvious sexual process.

In the more advanced group known as the Ascomycetes distinct male and female organs usually develop, but instead of spores being formed directly after fusion of the gametes, a number of filaments are produced which bear small club-shaped structures called asci. The contents of each ascus then round off to produce usually eight ascospores (literally spores produced in the ascus). Sexuality is even less obvious in the Basidiomycetes since no sexual organs are formed. Instead, spores are produced on a special structure called the basidium. As their mycelia grow, fusion between hyphae of different mycelia takes place with an exchange of nuclei. Each mycelium then passes daughter nuclei from cell to cell and when basidia are finally formed, the two nuclei inside each basidium fuse. This is equivalent to a sexual process and following it basidiospores (usually four) are budded off from the top of each basidium.

Within each major group of fungi some species are self-fertile and in others sexual reproduction can take place only when another fungus is present. Furthermore the two mycelia must belong to different or complementary strains. This phenomenon is called heterothallism. It was first investigated in the pin moulds but has been exten-

Above left
Coral Root Orchid (*Corallorhiza trifida*). This small yellowish European orchid is typical in having flowering stems arising from a cluster of swollen and deformed roots or mycorrhiza. Because the plant has no green leaves it is unable to live without this symbiotic mycorrhizal association.

Top
One of the simpler fungi, the typical spherical sporangia of the Pin Mould (*Mucor*) are of differing ages. When first formed the sporangia are colourless, but later turn black. When the sporangium wall breaks, a dome-shaped structure called the columella remains, with some of the spores still attached to it.

Above
When a spore germinates, the young hyphae radiate outwards, so that an undisturbed mycelium is circular. Here mycelia from several spores are growing under a log.

sively studied in the rust fungi. It is probably of widespread occurrence and is important in cross-breeding.

Wherever organic matter is present, providing there is sufficient moisture and a certain amount of warmth, fungi are likely to develop. Some fungi can tolerate quite extreme conditions such as marked changes of temperature. High temperatures do not inhibit growth unless there is a shortage of water. Other fungi can grow on rock surfaces, dissolving inorganic elements such as silica, iron and magnesium by producing citric acid. Dead wood is a widespread and important source of food for a number of fungi, particularly the larger mushrooms and toadstools as well as bracket fungi, as we can see if we walk through woodland, especially in autumn. Wood in buildings, and that stacked in timber yards may both be attacked. Of the known timber-rots the most serious is Dry

Rot Fungus which is very destructive in buildings throughout the cooler regions of the world. Smaller fungi can be seen at most times of the year growing on plant and animal remains such as fallen leaves and animal droppings.

Natural foods – cereals, fruit and vegetables–and manufactured foodstuffs, such as cheese, bread and preserves, will inevitably turn mouldy as they age, and spoilage of stored foods by both fungal and bacterial attack can cause considerable economic loss. Leather, paper and textiles are also attacked and this is particularly so in the tropics, where the warm moist climate encourages the growth of fungi overnight. Rubber, paints and varnishes, photographic films and even optical equipment can rapidly cloud over, the fungi often living on such restricted media as traces of oil and dust. In World War II this was a very serious problem in the Far East, where it was estimated that less than 50 per cent of supplies was fit for use. Some microscopic fungi live on hydrocarbon fuels, especially aviation fuel. As they grow, filters and pipes become blocked and electrical connections may be short-circuited. In modern box gir-

Top left
The Common Fairy Club (*Clavaria rugosa*) grows on woodland paths.

Bottom left
One of the jelly-fungi, the Sticky Coral (*Calocera viscosa*) resembles the fairy clubs (*Clavaria*) in shape, but has a slimy surface becoming horny when dry. Its golden-yellow forked fruitbodies are typically found clustered on conifer stumps.

Below
The Rust-spot Fungus (*Collybia maculata*) can be regarded as a typical gill-bearing Basidiomycete. It is common in woodland, its creamy white colour later becoming marked with reddish-brown spots.

Below
Trametes (Polystictus) versicolor is a small tiered bracket which may be found all the year round and causes much decay of felled timber. The upper surface is attractively zoned in shades of yellow, grey, brown or nearly black, the pores below being whitish.

Right
The hard hoof-like brackets of the Maze Fungus (*Daedalia quercina*) have irregular pores on the lower surface–hence the common name. It may be found on old oak stumps and lives for several years.

der bridges older anti-corrosive lead paints contained linseed oil which is broken down by some fungi, allowing damp and rust to penetrate. Special inorganic paints have now been developed to combat this fungal corrosion. Such deterioration and spoilage is expensive to combat but has to be regarded as part of the general process of decay.

Possibly even more serious are the diseases caused by parasitic fungi. Animals, including man, are less often attacked than plants and most of such fungus diseases are inconvenient rather than dangerous. They include the skin disorders ringworm and athlete's foot, but both farm workers and livestock may suffer serious lung damage by inhaling the spores produced on mouldy hay and straw. By contrast many of the plant diseases kill the host outright, often in a relatively short time. Field crops

such as cereals, potatoes, peas and beans, glasshouse and market-garden crops, as well as fruit trees and timber trees are all attacked by more or less specific parasites. Over the years losses of millions of pounds have resulted and sometimes, as in the case of potato blight, the course of history has been changed. It is estimated that fungi ruin or destroy at least 10 per cent of cultivated crop yields annually throughout the world. The single disease 'stem rust' of wheat caused crop losses of 23 per cent in the U.S.A. in the 1930s, and an outbreak in New South Wales, Australia in 1947–48 destroyed a year's potential food for 3 million people.

To combat fungal attack involves the combined efforts of many scientists for the fungus has to be identified and then the crop treated with appropriate fungicides. Fungicides range from the old and well-known remedies such as copper salts and sulphur, to modern organic preparations. Until recently these have been used as sprays but the newest techniques apply special fungicides which are actually taken up by the plant and spread throughout its tissues. Such systemic fungicides may be applied to the soil or to the leaves. Against such a catalogue of damage and loss it may well be wondered in what ways the fungi are of direct benefit to man. The list is of great significance. From the fungi we obtain a variety of foodstuffs and alcoholic drinks, antibiotics and other pharmaceuticals, and a range of industrial products such as organic acids and alcohol. Their direct use as food is increasing, and mushrooms are now cultivated in many parts of the world. However, it is the lower fungi whose metabolic activities are of such enormous significance to man, more especially the yeasts and blue-green moulds.

Yeasts occur naturally on the surface of leaves and flowers, commonly where sugary exudates are formed. Primitive man made use of these yeasts to produce crude fermented drinks and foodstuffs, some of which are still prepared today. The Japanese make a soup from fermented soya bean paste, and in Indonesia a similar paste is added to groundnuts, providing one-third of their protein requirements. Alcoholic beverages vary according to the source of carbohydrate to be fermented, thus temperate zones produce wine from grapes, beer and whisky from malted barley, and cider from apples. In Japan saki comes from fermented rice, in the West Indies molasses from the sugar cane is converted into rum. If a starchy base is used it must first be converted into sugars, as yeast cannot act directly on it. Sugar wastes, maize and potatoes are also used in the production of industrial alcohol.

ST. MARY'S ... GIRLS' SCHOOL
35 WEBER STREET WEST
KITCHENER

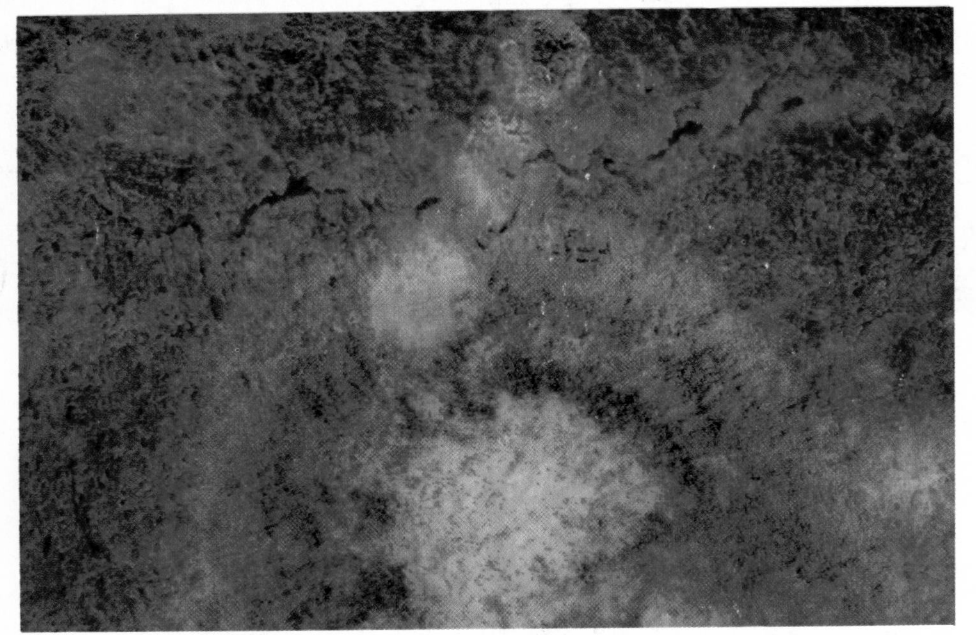

Above
The coral pustules of the Coral Spot (*Nectria cinnabarina*) are a familiar sight on fallen twigs and branches at almost all times of the year, and are evidence of gradual decay caused by the fungal hyphae. At first spores are formed, but these are followed in winter by dark-red warty flask-shaped fruitbodies, both stages being clearly shown in the photograph.

Left
A common contaminant in bakeries, Red Bread Mould (*Neurospora sitophylla*) can be a serious nuisance, its numerous spores spreading and ruining batches of bread. The photograph shows it together with other fungal colonies, mainly of *Penicillium*.

Below left
Military Club Fungus (*Cordyceps militaris*). Most members of this curious Ascomycete genus are parasites on insects, whose bodies they transform into a hard mass of hyphae. The photograph shows the reddish-orange club of one of the commonest species, growing on a moth pupa. The rough surface of the head is caused by the openings of innumerable small flask-shaped fruitbodies.

Above right
The small irregular crusts or brackets of the fungus *Stereum purpureum* have a hymenium which is lilac when young, grey when old. It may be found at any time of the year on plum, blackthorn and related trees. It is the cause of the serious disease known as 'silver leaf'.

In the process of fermentation carbon dioxide is given off. This is used in bread baking as the bubbles of gas lighten the dough.

Yeasts are used directly as a food source and are added, usually in dried form, to human food and also to cattle and poultry feedstuffs. More recently special strains of yeasts have been cultivated on a variety of bulk substrates. These include hydrocarbons from the distillation of oil, molasses, sawdust and wood or leaf pulp, and from them proteins or fats can be obtained, depending on the species of yeast. Yeast is extremely efficient at this type of conversion. From 100 kg (220 lb) of sugar or its equivalent, beef cattle produce 4 kg (9 lb) of protein, fowls 5 kg (11 lb), pigs 20 kg (44 lb), but yeast produces the surprisingly high figure of 65 kg (143 lb). Moreover, under ideal growing conditions some yeast species can quadruple their weight in 24 hours. There is no doubt that such food production techniques will become increasingly important and are rightly termed 'food engineering'.

The moulds *Aspergillus* and *Penicillium* are responsible for much deterioration and spoilage, but under controlled conditions provide a number of useful products. Certain strains of *Penicillium* are used in the production of cheeses and vitamins, as well as organic acids, are produced by large-scale deep-fermentation processes, using mainly species of *Aspergillus*. Citric acid, formerly obtained from lemons, is probably the best known of these acids and is used in the manufacture of soft drinks and some pharmaceuticals. Recently *Penicillium*-like fungi have been utilized to create vegetable protein rather as do some yeasts. A plentiful and cheap source of carbohydrates such as potatoes, yams, cassava or sugar cane, together with other nutrients, is converted into protein by controlled fermentation.

The most highly valued of all fermentation products, however, is undoubtedly penicillin, which was the first antibiotic to be isolated. The original species of *Penicillium* made famous by Sir Alexander Fleming was *P. notatum* but nowadays higher-yielding strains of *P. chrysogenum* are preferred. Some penicillins can be synthesized. A number of other antibiotics have subsequently been produced by different fungi or fungal-like species and together they have revolutionized the treatment of sores, infected wounds and long-standing scourges such as yaws, venereal disease and tuberculosis. It has been said that by producing antibiotics the fungi have more than compensated for any damage they cause, and mankind in general would uphold this verdict.

Major groups of fungi

Because the fungi are a complex group with a range of widely differing features, it is not surprising that there are varying systems of classification. However, all systems agree in dividing them into four major groups, excluding the lichens. Unfortunately the names of these groups and also of their subdivisions are cumbersome, and there are few common names for them.

Phycomycetes

The first and smallest group (about 1,400 species) is the Phycomycetes. Members of this group have a rather simple kind of mycelium, with branched hyphae lacking cross-walls. Some species lack a mycelium altogether. Many are moulds of one sort or another. They are the only fungal group to have non-sexually produced swimming spores (called zoospores), as well as sexually produced oospores, although they are not formed in all species. No member of the Phycomycetes forms large fruitbodies such as we find in the higher fungi, and there is no really significant feature (except perhaps the simple mycelium) which is common to all members.

A few examples will illustrate the diversity of the Phycomycetes. The water mould *Saprolegnia* is quite common on dead or dying fish in ponds and canals, but also in aquaria where it is often called 'gill rot'. Probably it does not attack healthy fish, but soon becomes established on injured or weakly specimens. It will also grow on other protein-rich matter such as animal droppings or decaying plant remains. Its branching hyphae produce elongate sporangia (spore cases) within which the zoospores are formed, and later sex organs. The zoospores can germinate immediately but the sexually produced oospores have a thick wall and undergo a resting period before eventually giving rise to zoospores.

Pythium is a genus of soil-living fungi which can also attack a variety of higher plants, particularly at the seedling stage when it causes the disease known as 'damping-off'. Anyone who has seen a tray of apparently healthy seedlings keel over within hours will appreciate the large-scale destruction which this disease can cause in plant nurseries. The hyphae attack at ground level, weakening the tissues of the host plant. Recovery rarely

The spores of the water mould known as Gill Rot or White Fungus (*Saprolegnia*) grow on weakly fish, developing into white patches of hyphae which choke the gills and rot the fins eventually killing the fish.

takes place but the disease can be controlled by watering with the fungicide 'cheshunt compound'. Damping-off spreads rapidly by means of zoospores but the more resistant oospores provide a reservoir of potential infection in untreated soils.

Phytophthora, the cause of Potato Blight, is a good example of a more specialized parasitic mould. Should spores drift on to the surface of a leaf they can germinate in a drop of rain or dew. From each spore develops a hypha which grows through one of the breathing pores on the leaf surface. Once inside the plant a mycelium forms, putting out small absorbing branches into the cells of the host plant. New spores are formed on special stalks which grow out through the stomata. Infected leaves at first have a fluffy white appearance but soon die, gradually weakening the plant. Fresh crops of spores, if washed into the soil, can infect the growing potato tubers and the whole plant quickly turns into a rotten mass, seriously reducing or even destroying the crop. Spraying with Bordeaux mixture, which contains copper salts, controls the disease. Burning of the potato haulms at the end of the season helps to reduce infection in the following year, since sexually produced spores can overwinter in the tissues. Other moulds related to *Phytophthora* may cause serious downy mildews on vines, hops, onions, lettuce and other crops.

The Pin Mould (*Mucor*) is a common saprophyte on decaying plant material and on manufactured foods such as cheese and bread. The related *Rhizopus* is found in similar places but may sometimes parasitize apples, causing a soft rot.

Ascomycetes

The Ascomycetes are the largest group with about 15,500 species. These vary in size and form, but rarely attain the size shown by the majority of the Basidiomycetes. They may be unicellular as in yeasts, or may have a well-developed mycelium of branched hyphae, with numerous cross-walls at more or less regular intervals which thus form cells. The hyphae commonly fuse with their neighbours so that a three-dimensional network is built up. Each cross-wall is perforated by a minute pore and so transport of essential food materials throughout the mycelium can readily take place.

The greenish-black fruitbodies of the Earth Tongue (*Geoglossum cookeianum*) may be found singly or in small tufts growing in grass, but because they are small are often overlooked. Ascospores are discharged from the enlarged head.

Members of the group have one characteristic feature in common. They produce spores known as ascospores (often in groups of eight) within a cylindrical or rounded sac–the ascus. Usually the asci develop grouped together within or on a definite fruitbody, and the form of this fruitbody is used to further divide the Ascomycetes. One subgroup, which includes the yeasts, has no fruitbody. The other, and very large subgroup (which has fruitbodies), has three main subdivisions. These are the Plectomycetes which have closed fruitbodies surrounding the asci; the Pyrenomycetes which have more or less flask-shaped fruitbodies and the Discomycetes which have open cup-like fruitbodies known as apothecia. Lining each apothecium is a fertile layer, the hymenium, in which the asci develop. A fourth group of Euascomycetidae is small and relatively unimportant.

As well as ascospores, many of the Ascomycetes develop other spores known as conidia and these may be more important in reproducing the species than are the sexually produced ascospores. Indeed, in some genera they have entirely taken over this function as in many of the blue-green moulds such as *Penicillium*.

We can consider selected examples from each of these groups in turn.

Hemiascomycetes

The yeasts have already been mentioned because of their extreme importance to man. The best-known genus is *Saccharomyces* and this and other species are found in a surprising variety of habitats. Although particularly abundant in flower nectar and on ripening fruits, both of which are rich in sugars, yeasts also grow on stems and leaves and in the soil. All yeast species are single celled, and ovoid or rounded in shape. They reproduce rapidly by budding and may sometimes form short chains of cells. Budding is most frequent when fermentation is most active, i.e. when the sugary medium is being converted into alcohol and numerous bubbles of carbon dioxide are given off. A few species multiply by dividing into two. Ascospores may be formed within the cells after a simple kind of sexual process, but this is not common.

Species of *Taphrina*, unlike the yeasts, have a normal mycelium. They are all parasites on the higher plants, particularly fruit and other deciduous trees. Spores germinate on the leaf surface and the hyphae grow through the epidermis into the leaf tissues where they absorb nourishment from the host cells. Eventually ranks of asci are produced just beneath the leaf surface and the ascospores are released when it ruptures. No proper fruitbody is formed. All species cause deformation either of the shoots, producing witches' brooms, or of the leaves by inducing the formation of brightly coloured blisters as in 'Peach Leaf Curl'. Such gross interference with leaf function saps the vitality of the tree and crop yields are markedly reduced.

Euascomycetes

The Plectomycetes produce their asci inside minute spherical closed fruitbodies, the ascospores being released only after the fruitbody wall has broken or decayed. The asci characteristically have a single-layered wall. Moulds such as *Aspergillus* (also known as *Eurotium*) may form fruitbodies, but the closely related *Penicillium* hardly ever does so.

In the growing season many different kinds of flowering plants show white dusty blotches of mildew on leaves, stems and flowers. These are the powdery mildews caused by the parasite *Erysiphe* and its relatives, probably the most well known of which is the Rose Mildew (*Sphaerotheca*). The powdery appearance is due to numerous small fruitbodies which bear variously shaped hairs. In humid conditions rose and other mildews spread rapidly forming vast quantities of spores. It has been calculated that from 1 cm^2 (0·155 in.^{2}) of infected rose leaf a half million spores can be produced.

Pyrenomycetes are often called flask-fungi because their small fruitbodies have a swollen base which encloses the asci, and a neck opening by a pore at the top. Through this pore the ripe ascospores are discharged in a drop of slime. Such a fruitbody is called a perithecium. Dutch Elm disease (*Ceratocystis ulmi*) produces particularly long-necked perithecia beneath the bark of elm trees killed by the fungus. Some flask-fungi produce large numbers of perithecia embedded in a firm mass of sterile hyphae. This mass is called a stroma and has a definite shape depending on the species. The aptly named King Alfred's Cakes (*Daldinia concentrica*) grows saprophytically on dead trees such as ash, beech or birch and forms black stromata about the size and shape of a currant bun. If cut open these are seen to consist of alternating dark and light concentric zones. Small perithecia containing numerous asci develop just beneath the surface and from them small black ascospores are actively discharged during the summer months. The Candle Wick Fungus (*Xylaria hypoxylon*) has black ir-

Left
The Doughnut Fungus (*Rhizina undulata*) is a flattened convoluted Ascomycete attached to the soil or wood on which it grows by many stout fibrils. It is common after fires, especially in conifer woods where it is a pest, attacking the roots of pine seedlings.

Right
King Alfred's Cakes or Cramp Balls (*Daldinia concentrica*). Either name is appropriate for these rounded black fruitbodies. The first is obvious, but the name Cramp Ball refers to an old belief in its value as a charm. It produces numbers of ascospores in wet or dry weather, usually only at night.

Below
The Candle-wick Fungus (*Xylaria (Xylosphaera) hypoxylon*) is a small Ascomycete which grows on old logs. It appears as small simple or branched clubs which are black below and powdery white above. This is the spore-bearing or conidial stage; when the fruitbodies develop the whole club turns black.

regular elongate stromata about 5 cm (about 2 in.) high. These are coloured with white powder (due to conidia) when young, but are dark with perithecia when mature.

The Discomycetes are a large subgroup with over 4,000 species. Their ascocarps usually take the form of an open cup when mature, with the asci densely packed in a fertile layer (the hymenium). Surprisingly the truffles are grouped here and can be related to more typical forms by intermediate types.

The best known genus is the Elf-cup (*Peziza*). Species vary in size from the minute scarlet cups of the Woolly Disc Fungus (*P. scutellata*) which measure only 1 cm (0·4 in.) across, to large specimens of the well-named Orange Elf-cup or Orange Peel Fungus (*P. aurantia*) which may be as much as 20 cm (about 8 in.) in diameter. The genus is common and widespread in temperate regions, growing on the ground in woodlands, but with some species associated with richer soils such as compost heaps.

The morels (*Morchella*) have a fruitbody of two distinct parts: a stout hollow stalk and a club-shaped head with a highly folded and wrinkled hymenium. The appearance is that of a greyish or brownish sponge and at first sight there seems little relationship to the open cup of *Peziza*. Nevertheless there are a number of intermediate types of Discomycete which link the two.

Discomycete asci open either by a lid or by ejecting a plug, and in both types pressure builds up so that the ascospores are actually shot out in rapid succession one by one. If a mature cap is touched thousands of spores may be simultaneously ejected giving the effect of a puff of smoke.

Above
In the Scarlet Elf-cup (*Peziza coccinea*) the fruitbodies are stalked, and this shows clearly in younger specimens. At first they are closed but as they mature the fruitbodies become almost flattened.

Right
The Orange Elf-cup or Orange Peel Fungus (*Peziza aurantia*) bears bright orange fruitbodies which are always conspicuous and often appear on bare soil. Asci are borne on the inner surface.

Basidiomycetes

The third major group comprises the Basidiomycetes. With 15,000 species this ranks second in size to the Ascomycetes and includes such familiar fungi as the mushrooms, bracket fungi and puff balls. A number of less conspicuous species also belong here, together with two important parasitic groups—the rusts and smuts.

The Basidiomycete mycelium has many features in common with that of the Ascomycetes. It too has branched hyphae with perforate cross-walls at more or less regular intervals, and with the hyphae fusing to form a network. But whereas the cells of the Ascomycete mycelium contain one to several nuclei, most of the Basidiomycetes have two. Furthermore there is usually a characteristic swelling at each cross-wall. These structures are known as 'clamp connections' and they provide a passageway through which nuclei can migrate from cell to cell. However, the fundamental characteristic of the Basidiomycetes is the formation of basidiospores. These spores bud from club-shaped structures known as basidia which are usually grouped together in a fertile layer—the hymenium. In the higher Basidiomycetes the hymenium develops either within a fruitbody, as in the puff balls, or on the outside. The hymenium may cover gills or folds or may line narrow tubes; all these types are shown by the bracket and cap-and-stalk fungi, but other forms of fruitbody are known.

The Basidiomycetes are divided into two main groups: the Heterobasidiomycetes which include the smuts, rusts and jelly-fungi, and the Homobasidiomycetes which include almost all of the better-known fungi. These are the mushrooms and toadstools of woods and pastures, puff balls and earth stars, the nauseating stinkhorns and the bracket fungi found on both old logs and living trees.

Heterobasidiomycetes

The smuts and rusts are all parasites on higher plants and a number of them cause crop diseases of major economic importance. The smuts are so called because of the black dusty spore masses they produce. Different species can be found in all parts of the world, attacking cereals and other crop plants. The black spores are not basidiospores, but resting spores which give rise to basidiospores on germinating in the soil. The infective mycelium is then produced and attacks nearby host plants. In cereals the smut spores are produced in the ears of grain, the seed failing to develop, and in place of each grain as many as 5,000 spores may form. At the end of the growing season, depending on the species, ripe spores in their millions are either blown away, or released when the crop is harvested.

The rusts are a large group of several thousand species and they cause even greater losses in a wide range of cultivated plants. Cereals, peas and beans, apples and other fruits, coffee, and timber trees (particularly pines) are among those affected. Glasshouse plants may also succumb to attack, for example, the commonly grown carnation. Probably the most important and therefore the most studied is *Puccinia graminis*, which causes Stem Rust or Black Rust of wheat and other cereals and grasses.

Like many, but not all rusts, *P. graminis* has a highly complex life cycle involving the production of five kinds of spore and the infection of two host plants, the wheat (or other cereal) plant itself and the common barberry. In the spring

resting spores give rise to basidiospores which invade the leaves of the barberry plant and grow into a mycelium. Infected leaves develop small galls, in some of which chains of spores are produced; these can infect wheat or other cereal plants. Infected wheat plants produce enormous numbers of yet more spores, of a different kind known as summer spores, and as the season progresses these spores are blown from one wheat plant to another. The stems and leaves become covered with powdery orange blotches giving the characteristic 'rusted' appearance to the weakened plants. Shortly before harvest, black patches develop and these contain dark-coloured resting or winter spores which fall to the ground. In the following spring they give rise to the basidiospores which infect the barberry, and complete the cycle.

Stem rust is a very important cereal disease in Australia, where the highly infective spores can survive the winter in the soil, and also in North America where reinfection takes place from spores produced earlier in the year in the far south. Vast clouds of spores are blown northwards to the wheat belt and this source of infection may possibly be supplemented by the northward migration of large flocks of birds carrying the spores on their feet and plumage. In Britain stem rust is of lesser importance since the alternative host, the barberry, is rare and the winters are too cold for the spores to survive. Elsewhere the disease is of major importance with crop losses of thousands, sometimes millions, of tons of wheat. New genetic strains of wheat rust periodically develop, usually originating in northern Mexico or the southern states of America, and to date some 200 strains have been identified. To counteract this the plant breeder is constantly trying to produce new resistant wheat varieties and thus there is a continuing battle between the invasive properties of the parasite and the resistance of the wheat plant.

The jelly-fungi comprise the third group of Heterobasidiomycetes and all have open fruitbodies of soft, gelatinous or cartilaginous texture. For this reason they may also be classified with the cap-and-stalk and bracket fungi. The best-known example is the Ear Fungus (*Auricularia*) and this and related genera are widespread in temperate and tropical regions of the world.

Left
This field of ripening barley shows infected and uninfected ears of grain. The characteristic black resting spores of Barley Smut (*Ustilago hordei*) take the place of the developing seeds, and may infect healthy grain during harvesting.

Above
The bright gelatinous fruitbodies of the Yellow Brain or Orange Jelly Fungus (*Tremella mesenterica*) are common at any time of the year, growing in highly folded masses on dead wood.

Homobasidiomycetes
The higher Basidiomycetes are known as Homobasidiomycetes and have sizeable fruitbodies producing club-shaped basidia without cross-walls, thus differing from Heterobasidiomycetes. There are two subgroups. Those in which the hymenium is enclosed until after the spores have ripened form the far smaller subgroup known as the Gasteromycetes. Fungi which have their basidia in a fertile layer which is exposed when mature are known as Hymenomycetes; their spores are shot off when ripe. A few examples will serve to demonstrate the range of form of both subgroups.

Puff balls are the best known of the Gasteromycetes, most species belonging to the genus *Lycoperdon*. They have roughly pear-shaped fruitbodies with a papery wall which develops an apical pore when the spores are ripe. These spores are minute in size and are formed in

Top left
Singly or clustered, the warty greyish Common Puff Ball (*Lycoperdon perlatum*) appears in grassy places. If lightly touched, for instance by raindrops, *Lycoperdon perlatum* emits small puffs of spores from an obvious apical pore.

Bottom left
An average-sized Giant Puff Ball (*Lycoperdon (Calvatia) giganteum*) is about 30 cm (1 ft) across but the largest known specimen, found in New York State in 1877, measured well over 1 m (over 3 ft) wide and almost $1\frac{2}{3}$ m ($5\frac{1}{2}$ ft) long. At a distance it was mistaken for a sheep! Spore production is enormous; ripe specimens puff out thousands of millions of spores through splits in the fruitbody wall during a period of several days.

Top right
In the Earth Star (*Geastrum triplex*) the spore mass is enclosed by a delicate wall, which is itself surrounded by the fruitbody wall. When ripe, the fruitbody wall ruptures as shown here.

Bottom right
The fruitbody wall of *Geastrum triplex* finally folds right back so that the spore mass, within its case, appears to perch on top. A slight touch, as from a rain drop, is sufficient to discharge a cloud of spores.

large numbers. If the puff ball is touched, or should raindrops fall on it, a cloud of spores can be seen escaping through the pore.

The Giant Puff Ball (*Lycoperdon* or *Calvatia gigantea*) grows at least as large as a man's head and often exceeds this. Calculations on the number of spores formed by such a fruitbody produce some astonishing figures; an average specimen can give rise to 7 billion spores and a really large specimen might produce 100 trillion spores! It has been estimated that if each of these spores could grow into a mature fungus, the resultant mass of living matter would be nearly one thousand times the volume of our planet.

Earth stars (*Geastrum*) are woodland fungi closely related to puff balls but much less common. Both look alike when young but as the earth stars mature their two-layered wall separates. Then the

Above
The fruitbody of the Stinkhorn (*Phallus impudicus*) arises from an 'egg'. One here is seen bursting through the ruptured wall. Also shown here are the funnel-shaped caps of another Basidiomycete—*Lentinellus cochleatus*.

Right
A mature fruitbody of the Stinkhorn (*Phallus impudicus*), its cap covered in a foetid dark spore-containing slime which is dispersed by flies.

outer layer splits into four or more rays with the inner layer, still surrounding the spores, perched on top.

When walking through woodland in late summer and autumn one may sometimes be aware of a foul odour—somewhat like that of decaying flesh. By tracking down its source one may find one or more ripe stinkhorns surrounded by a milling mass of flies. Varieties of stinkhorn are more common in warmer regions, but some species of *Dictyophora* are found in North America, northern Europe and Japan. These have a delicate lacy frill around the base of the cap. Better known, especially in northern Europe, is *Phallus impudicus*. All stinkhorns develop from egg-like structures produced by white rhizomorphs growing in the dead leaves on the floor of the wood. Within the 'egg' the basidiospores ripen in a slimy mass and then, within a few hours, are borne aloft by the rapid growth of a stout stalk. The slimy mass of spores exudes the foetid odour so attractive to flies which then eat both spores and slime. The spores pass unharmed through the fly and are excreted elsewhere. Since each excretion can contain over 20 million spores, the potential dispersal of the spores is enormous.

Yet another method of spore dispersal is shown by the Gasteromycete known as *Sphaerobolus*, which has a catapult mechanism. The spores develop inside a many layered cup. Pressures build up between the layers and the cup suddenly turns inside out, throwing the spore mass for a distance of 5–6 metres (16–19½ ft).

Below
The basket shape of the tiny and intriguing Bird's Nest Fungus (*Crucibulum vulgare (laeve)*) with its basidiospore-containing 'eggs' is here clearly shown. When young the fruitbody is closed by a lid. The eggs are subsequently dispersed by raindrops.

Bottom
Dryad's Saddle (*Polyporus squamosus*) is a conspicuously large bracket fungus easily recognized by its scaly upper surface and saddle shape. It attacks living trees, causing a white rot, and also lives on the dead wood. Spore production is enormous; a recorded specimen produced 11,000 million spores in a period of 13 days.

Still more curious are the bird's nest fungi (*Cyathus* and *Crucibulum*). These have small basket-like fruitbodies (up to 2 cm or about $\frac{3}{4}$ in. across) with a lid which dries up to reveal tiny egg-like spore masses called peridioles. When it rains the drops of water splash the 'eggs' out of the cup and up to 2 m ($6\frac{1}{2}$ ft) away. The 'eggs' each have a long attaching thread which suddenly snaps at the base, entangling the spore mass in nearby vegetation.

An open fruitbody characterizes the Hymenomycetes, and this group includes such familiar types as mushrooms, toadstools and bracket fungi. There are two sub-groups: the polypores which include the brackets, clavarias and tooth fungi, and the agarics which are cap-and-stalk fungi.

Of the two groups the polypore fruitbody is far the most variable. It may be a thin or thick crust spreading over the surface of log, tree or other source or timber, and this is well shown by the Dry Rot Fungus (*Serpula* or *Merulius lacrymans*). Dry rot and other timber rots have been known since earliest times and in early writings it was suggested that timber should be treated with various oils or with pitch. *Serpula* is a fungus of domestic, farm and other buildings and is said to occur wild only in the Himalayas. It is essentially a fungus of temperate, even sub-arctic, conditions and does not flourish in the tropics. The name 'dry rot' is misleading since the spores require moisture to germinate and dry, well-seasoned timber is rarely affected. The specific name *lacrymans* refers to the drops of moisture produced as the rot begins to take hold. When the mycelium is established it sends out extremely strong rhizomorphs which can spread through dry material such as brickwork to attack previously uninfected wood, as well as organic materials such as paper and leather. Signs of wood rot include a very characteristic, somewhat fishy smell, and discoloration and softening of the affected timber, which later cracks and turns powdery. When fruitbodies are finally formed they appear as white-margined masses about 1 cm (almost $\frac{1}{2}$ in.) thick and varying from a few centimetres to a metre or more (over 3 ft) across. The surface is honeycombed; this is the hymenium which produces cinnamon-orange spores in vast quantities. These drift on air currents and thus spread the infection.

The tooth fungi include all species which have a hymenium in the form of spines or teeth on the lower side of the fruitbody. *Hydnum* resembles a cap-and-stalk fungus, but the difference is obvious if we examine the under surface of the cap. The related *Hericium* includes some of the most beautiful of all fungi. *Hericium coralloides* is widespread in north temperate regions but is never common. It has a much branched fruitbody, white and almost feathery in appearance, with long pendant teeth.

The delicious Chanterelle (*Cantharellus cibarius*) and Horn of

Plenty (*Cratarellus cornucopioides*) resemble agarics except that the 'gills' on the lower surface of their deeply funnel-shaped caps are really irregular folds. But the most typical of the polypores are the bracket fungi. These produce the shelf-like or plate-like fruitbodies so obvious on standing or fallen tree trunks. They are leathery or corky in texture, lasting for several months. A few live for a number of years. The lower surface of the bracket shows numerous tiny pores and these are the openings of small parallel tubes which are lined by the spore-producing hymenium. Many attack living trees, the spores entering through a wound, and the mycelium grows through the trunk and branches causing destructive rots. While the young fruitbody is developing it responds to the force of gravity in such a way that the tubes grow precisely vertical. The basidia are horizontal and the mature spores are shot into the centre of the tubes, then fall clear of the bracket and drift away in air-currents. Most of the leathery or corky brackets can endure dry conditions for weeks or even months, although spore discharge ceases. In moist weather the dry brackets absorb water, and spore production is resumed. One of the more common annual species is the Dryad's Saddle (*Polyporus squamosus*) which grows on a number of deciduous trees, especially elm, and can cause a serious heart rot. Perennial brackets become very woody,

Grifola (Polyporus) gigantea usually grows in groups on oak or beech, and may be up to 1 m (39 in.) across. The upper surface of this common fungus is mid-brown with whitish pores below, and these and the flesh turn almost black when bruised.

Left
Tinderfungus (*Ganoderma applanatum*). Before matches were invented, pieces of corky brackets such as *Ganoderma* and *Piptoporus betulinus* were kept alight but smouldering in closed tins ready for fire lighting.

some surviving for many years. For example, *Phellinus* or *Fomes igniarius* grows throughout the north temperate zone on poplar and willow, and actively produces spores for twenty to thirty years. In America some specimens are known to be at least eighty years old. The tubes increase in length, but the older parts gradually become blocked and cease to function, only the youngest parts (two or three years old) producing spores.

Ganoderma also has perennial brackets but these rarely last more than ten years. The well-known and cosmopolitan species *G. applanatum* forms thick woody brackets as much as 60 cm (2 ft) across. It is especially common on beech trees and causes a serious heart rot. Between spring and autumn the bracket actively produces spores, and a large specimen can release 20 million spores a minute. Since spores are produced daily for about five months in the year, the total number of spores liberated during its lifetime is truly astronomical.

The agarics are a large group of some 7,000 species and include gill-bearing cap-and-stalk fungi (although a few look rather bracket-like), and the boleti. Species of *Boletus* vary in colour and size, but all have a pronounced and fleshy cap with numerous pores beneath, which lead into hymenium-lined tubes. Almost all are free-living, but *B. parasiticus* attacks earth balls (*Scleroderma*), producing tufts of fruitbodies on them.

The gill-fungi have been among the most investigated of all groups of fungi. Although they show differences in size, colour and texture, the important differences lie in the development of the fruitbody, the form of the gills and the colour of the spores. This last point is easily determined by making a spore print. The cap is cut off and placed gills downward on a piece of paper, pale grey for choice, and left for a few hours in a draught-free place. If the cap is then carefully removed, radiating lines will show the number and arrangement of the gills and the colour of their spores.

Bottom left
Fir Murderer *(Fomes annosus)* is the name of this highly destructive parasite of conifers, which also grows on pit props and other structural timbers. The brown upper surface blackens with age, the margin and lower surface remaining white. Related species attack other types of tree.

Above
All boleti, such as *Boletus bovinus*, have pores instead of gills on the under surface of the cap, and here they are well shown.

Right
The only parasitic species of *Boletus* is *Boletus parasiticus*. This characteristically attacks the Common Earth Ball (*Scleroderma aurantium*).

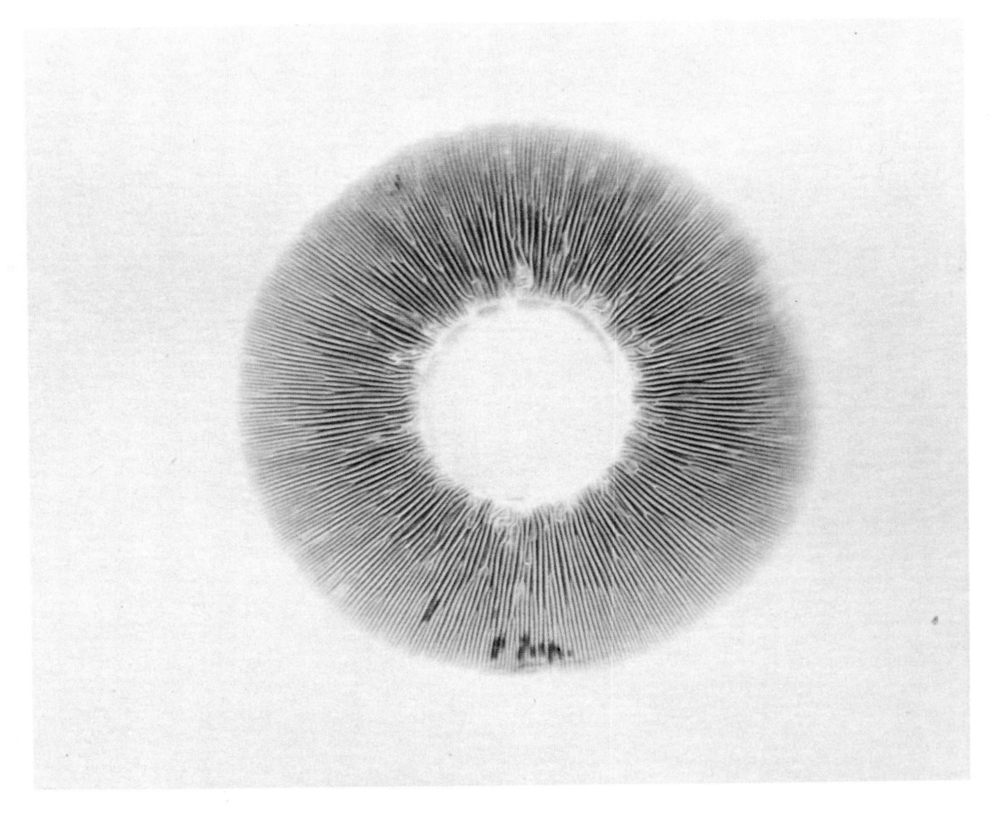

Right
A spore print of the Cultivated Mushroom (*Agaricus bisporus*) shows the number, position and varying length of the gills. The clear central space locates the stalk. Because spore production is so copious, a cap placed gills downward on paper will overnight produce a very distinct outline of the gills.

Below
The common tufted Scaly Cluster Fungus (*Pholiota squarrosa*) grows on deciduous trees. It is thickly covered with shaggy mid-brown scales and produces brown spores.

About one-third of all agarics have white spores, for instance *Amanita, Tricholoma* and *Russula*; one-quarter have brown spores, for instance *Pholiota*; and the remainder have pink, for instance *Entoloma*; purple as in the mushrooms (*Agaricus*); or black spores as in the ink caps (*Coprinus*). The gills are arranged like the spokes of a wheel from the stalk to the edge of the cap, but with some short gills in between. Gills are wedge shaped in section and hang vertically downwards beneath the cap, but their attachment near the stalk differs according to the species. In some, for instance the mushroom, the gills hang down free, in others, such as *Russula*, they are attached in the angle between cap and stalk. Some fungi, for instance *Hygrophorus*, have the gills running part-way down the stalk and a few have a depression or wavy edge near the stalks, as seen in some species of *Cortinarius*.

The number of gills varies. In a Field Mushroom with an 8–10 cm (3–4 in.) cap there may be over 400 gills. The hymenium covers the entire gill surface and so forms a far greater area than if the cap were flat underneath. It has been calculated that a cap 10 cm (4 in.) across has a hymenial surface of 1,235 cm^2 (1·3 ft^2) and thus spore production can be greatly increased.

The fruitbodies develop from small knot-like bodies on the my-

celium. As these enlarge, the distinction into cap and supporting stalk becomes apparent. In some species the gills are uncovered throughout their development, but in most species the gills are protected when young by a thin membrane known as the 'partial veil' or velum. When this finally tears, exposing the gills, it may disappear completely, or a remnant may remain, usually as a ring or annulus on the stalk. In the genus *Cortinarius* the young gills are protected by a delicate cobwebby veil which remains as a kind of frill hanging down from the edge of the caps. The familiar cultivated mushroom clearly shows both partial veil and ring as the button stage enlarges. In some species of fungi the developing fruitbody is totally enclosed by a 'universal veil'. This enveloping membrane is exhibited by the highly poisonous genus *Amanita*, and when it is split by the growing fruitbody, it forms a sac-like structure (the volva) at the base of the stalk. Remnants may also remain as scales or warts on the cap – as shown by the white-spotted red cap of the Fly Agaric (*A. muscaria*).

The enveloping membrane so characteristic of Amanitas, and which is left usually as a sheath (volva) at the base of the stalk, shows in this young Panther Cap (*Amanita pantherina*). The white spots on the cap are also remnants. The low-set ring on the stalk is the torn membrane which covered the young gills. This is a highly poisonous species and should be avoided.

The basidiospores take eight hours to form, but their discharge from the basidium takes under a minute. To be successful the process must be very precise since the space between adjacent gills is only a fraction of a millimetre. One by one, in rapid succession, the ripe spores are violently shot from the basidium, each with a small drop of water attached to it. The spores are shot horizontally for a minute distance, approximately midway between the gills, and then fall clear of the cap by the weight of each water drop. For this to be successful the gills must be absolutely vertical. Small movements of both stalk and gills are made in response to gravity, and this ensures accurate positioning of the gills and the successful release of the spores. Once free of the cap, the minute water drop on each spore evaporates and the dust-like spores are carried away by air currents.

The actual number of spores produced by an agaric depends on the life of the fruitbody as well as its size. Some minute species of ink cap may last barely a day. The average active life is probably two to three days but some of the stouter species may last five to six days. Mushrooms may last nearly a week; like all agarics they discharge their spores day and night. We can calculate that a 10 cm (4 in.) mushroom cap would, during its life, produce 16,000 million spores and these would be discharged at an average rate of over 100 million spores an hour.

The ink caps (*Coprinus*) are easily recognized by their tall, slender whitish caps with thin parallel-sided gills. The gills age from the bottom, changing from white to black as the spores ripen. Special hairs act as 'spacers' and keep the narrow gills apart. The older parts of the gills undergo autodigestion and turn black and liquid, falling away in inky drops—hence the common name.

In woodlands in autumn we may notice that some of the fungi appear to grow most often beneath particular trees. One of the more obvious examples is the Fly Agaric (*Amanita muscaria*) with its white-spotted red caps, commonly found beneath birch trees, or less often under conifers. The Saffron Milk Cap (*Lactarius deliciosus*) grows under pine trees, as do several species of *Boletus*. In North America the brownish-yellow *B. americanus*, and in Europe the similarly coloured *B. bovinus*, are common in pine woods. Some boleti are more specific; *B. edulis* for example, is almost always associated with larch. Beech trees are associated with some species of *Russula*, for instance *R. velenovskyi* with edible yellowish-red caps, and *R. maculata* which has flesh-coloured caps with darker spots. Although such pairings may be obvious it is more common for a number of different fungi to associate with several other tree partners. Research has shown that the Scots pine is capable of forming associations with well over one hundred species of Basidiomycete. Further, one tree can partner several different fungi at the same time.

Such associations are termed mycorrhizal, and they have been the subject of investigation for a number of years. Each involves a definite relationship between the fungus and the tree since the fungal

Far left
These young specimens of the Shaggy Ink Cap or Lawyer's Wig (*Coprinus comatus*) show the shaggy, almost woolly scales which give the fungus its common name.

Left
These mature caps of *Coprinus comatus* are undergoing autodigestion, with older parts of the gills falling away in inky drops and leaving the younger parts free to disperse their spores.

Above
This curious-looking Ink Cap (*Coprinus lagopus*) is several days old and in this species, as the gills age they roll upwards, so that there is no interference with spore release from the younger parts of the gills. Autodigestion of the gills, seen in most other ink caps, does not take place.

Top
Giant Clitocybe (*Clitocybe (Leucopaxillus) gigantea*). The gills of all species of *Clitocybe* run part-way down the stalk. This species is one of the largest cap-and-stalk fungi of temperate regions with a cap up to 30 cm (1 ft) across.

Above
The Sulphur Tuft (*Hypholoma fasciculare*) is an easy fungus to recognize because of its sulphur yellow cap and stalk, and its habit of growing in large tufts on tree stumps.

mycelium, as it grows, actually infects the smaller tree roots. These soon become distorted, with a characteristic swollen and much-branched appearance. Such fungus-root structures are termed mycorrhizas. Each is enclosed in a dense sheath of hyphae, some of which penetrate between the outer cells of the root. It is assumed by most mycologists that the association is a symbiotic one, in other words, of mutual benefit to both the organisms concerned. The tree stores manufactured starch and sugars in the roots and these can be absorbed by the fungus. This in turn is particularly efficient at absorbing mineral salts from the soil and presumably the fungal component passes the salts on to the cells of the tree roots. Since there may be a shortage of these salts in forest soils, especially nitrates, such transfer is extremely valuable to trees. Over fifty years ago it was shown that seedlings of pine and spruce would grow normally in poor soils only when they had developed mycorrhizas, and uninfected seedlings died. It has actually proved possible to induce mycorrhizal roots in pine seedlings by adding mycelium from species of *Boletus, Amanita* and other fungi.

The type of mycorrhiza found in forest trees is known as ectotrophic, and the contributing fungus seems always to be one of the larger Basidiomycetes. Orchids and some saprophytic plants also form mycorrhizal associations but of a different type. Here the fungus is usually one of the smaller Ascomycetes (less commonly a Basidiomycete) and it penetrates the outer root cells but does not form an outside sheath. This type of mycorrhiza is termed endotrophic. It is important for the well being of the plant, and indeed, orchid seeds are unable to germinate if the fungus is absent. Endotrophic mycorrhizas are also found in many annual and perennial herbaceous plants, but have been less well studied. The infecting fungi may be species of Phycomycetes and include such soil-living types as *Pythium.*

Mention was made earlier of Ascomycetes in which asexual spores are produced, but not fruit-bodies. For the sake of convenience many mycologists place these fungi in a fourth and special group, the Fungi Imperfecti (literally 'incomplete fungi'). Normally these do not reproduce sexually and they constitute a surprisingly large group of well over 10,000 species. They grow on a wide range of plant materials including ripe fruit and vegetables, manufactured foodstuffs, paper, textiles and leather, and are responsible for considerable losses by contamination and spoilage. They may also live in the soil and more than half of all known species of soil fungi belong to this group, many being species of *Penicillium* and *Aspergillus*. Some Fungi Imperfecti cause diseases in plants or animals and in varying ways many species make a considerable impact on man.

The mycelium is formed by the germinating spore and quickly produces spores which in large numbers may colour the mycelium black, grey, blue-green or yellow. These spores are readily detached and, depending on the species, may either be blown away or may become covered by slime and subsequently washed away by raindrops. *Penicillium* and *Aspergillus* are two closely related genera but they show differences in the way in which their air-dispersed spores are borne. In *Penicillium* these develop in chains from the tips of a number of parallel branches borne on a slender stalk. *Aspergillus* also produces spores in chains, but they radiate outwards from a stalked swollen head.

Below
The pretty little Fan (*Schizophyllum commune*) can be found on tree trunks for most of the year. There is no stalk, and the gills are peculiar because they are split vertically along their edges. Spores are released only in damp weather. When dry the spores are protected by rolling of the half-gills.

Bottom
Porcelain Fungus (*Oudemansiella mucida*). The common name is obvious if we look at the photograph showing one cap seen through the almost transparent cap and gills of another.

Right
One of the best known and widely distributed common encrusting lichens, *Xanthoria parietina* grows in many different habitats including rocks, walls and roofs. It grows best by the sea or on dusty farm buildings, but is here shown on an old tombstone, together with small patches of a grey encrusting lichen.

Below
In clean air, lichen growths on trees are abundant. Here several species of lichens are growing together, of which the largest are the much-branched and hair-like beard lichens (*Usnea* and *Alectoria*).

Botrytis is a common and widespread grey mould which grows both as a saprophyte or as a parasite. It flourishes in warm moist conditions and is common on ripe strawberries and other soft fruits. The spores are borne on side-branches and look like microscopic bunches of grapes. These are also air-dispersed but the fungus is also spread, as well as over-wintered, by sclerotia. The genus *Fusarium* is an important cause of wilts of crop plants such as potatoes and bananas because the mycelium blocks the conducting tissues. The spores are dispersed in slime.

Such a brief dismissal of so important a group as the Fungi Imperfecti does not reflect their true significance to man. As a group there is no doubt that they include some of the most useful, and at the same time some of the most destructive of all fungi.

Lichens

The lichens are not true fungi, but because of their body structure, which is unlike that of any other plant group, they deserve a brief mention. Lichens are sometimes called dual organisms, because the plant body is formed by an extremely close association between a fungus and an alga. This takes the form of a 'sandwich', with a middle layer of loosely woven hyphae in which the algal cells are embedded, between layers of compactly arranged hyphae.

It is generally considered that this association is another example of symbiosis, since the fungus lives by absorbing food manufactured by the alga, which in turn benefits in several ways. It is supplied by the fungus with essential inorganic substances and it is protected from the dangers of drying out or of excessive sunlight, which can damage the algal chlorophyll. The fungus concerned is most commonly an Ascomycete—certainly so in Europe—but may sometimes be a Basidiomycete. The algal constituent is either a simple green alga or a member of the blue-green algae.

Lichens form a large group with over 16,000 known species and these grow in a wide range of habitats, some of them seemingly impossible for plant growth. Thus we may find lichens on rock surfaces not only near the North and South Poles but also in deserts where rainfall is minimal and temperatures extreme. In the Negev in Israel, species of the greyish tufted lichen *Ramalina* tolerate daytime temperatures of over 80°C (176°F) and a dramatic fall to below freezing point at night. Another species of *Ramalina*, together with other lichens, grows on coastal rocks splashed by salt spray. Many lichens grow on dry heathlands and poor or even bare soils. A number grow on tree trunks and some species may form festoons several metres long.

Mosses and lichens are the first plants to colonize bare soils. The recently formed volcanic island of Surtsey (off the coast of Iceland) is showing early signs of colonization by lichens. The only condition which lichens cannot tolerate is atmospheric pollution. This is probably because their rate of growth is so extremely slow that any factor which interferes with their metabolism kills them. For this reason several species have died out within the past 100 years. All species are perennial, and measurements of growth show an annual increase of a few millimetres only. By simple calculation we know that some undisturbed colonies are more than 2,000 years old.

There are three groups of lichens differing in their growth form. One group includes species which are flat and encrusting, another group is flattened and leaflike, and the most advanced group has a bushy, almost twiggy form. In colour they are commonly greyish-green but may be black or even bright orange. The familiar orange-yellow patches on farmyard roofs and on seaside boulders is the lichen *Xanthoria*.

Lichens reproduce by budding off small clusters of algal cells surrounded by a few fungal hyphae. Cup-shaped apothecia may also form, which, as in the Discomycetes, bear asci on the inner surface. The hymenium may be green or brightly coloured—it is red in *Cladonia coccifera* for example.

Lichens are of little general importance but were a former source of vegetable dyes. The chemical indicator litmus is also a lichen product, and a few species are still used in the making of perfumes. Some species produce antibiotic substances, but these are unlikely to be exploited commercially. In the tundra and similar subarctic wastes, however, the lichens will always be essential as fodder for reindeer and will thus continue to form a substantial part of the local economy.

Common Cup Lichen (*Cladonia pyxidata*). Cladonias have a semi-prostrate body usually growing on soil, many being found on peaty moorlands. This species is abundant and cosmopolitan and produces small, stalked powdery cup-like fruitbodies.

Edible fungi

With its brown bun-shaped cap and squat swollen stalk, the Cèpe (*Boletus edulis*) is an easy species to recognize. It has an excellent flavour and should be picked while the tubes are white or yellowish. These are removed before cooking.

The use of group names for edible and non-edible fungi is not very precise. Some authorities use 'mushrooms' as a general term for all edible species, and fungi or 'toadstools' for the remainder. Others restrict the term mushroom to the genus *Agaricus* together with a very few other edible species. In this book the more general use of 'mushroom' will be employed. The derivation of the term 'toadstool' is precisely what the name suggests and dates roughly from the Middle Ages. Brittany uses a somewhat similar name meaning toad's hat or bonnet.

Of the many fungi found in different parts of the world some at least would have been eaten by early man. In biblical times mention is made of fungal diseases of cereal crops, and the use of yeast (leaven) was understood, but there are no references to edible fungi as such. We know from letter tablets (c. 1800 BC) found in what is now Iran, that the inhabitants of this eastern region of the Mediterranean recognized several poisonous as well as edible species. Truffles were highly valued and they are still abundant in that area.

There are numerous references to fungi in the writings of Greek and Latin authors. Greece was, and still is, very poor in natural fungi and the wealthier Greeks had to import mushrooms as a luxury food from Italy. Despite this the Greeks seem, on the whole, not to have been keen on them. By contrast, the Romans were both enthusiastic and knowledgeable about edible fungi. Upper class Romans were very fond of certain fungi, the most popular being *Amanita caesarea*. This is the highly esteemed 'egg mushroom' of modern Italy and the Imperial Mushroom of ancient Rome. Somewhat confusingly the Romans called it *boletus* and hence the special cooking vessels used to heat the mushrooms were known as *boletaria*.

Next to *A. caesarea* in popularity was *Boletus edulis* (now known to the French as the Cèpe and to the Italians as Porcini, and also truffles.

The Romans called truffles *tubera* and imported them in large quantities from Libya and other parts of North Africa. Pliny (23–79 AD), a great student of natural history, discussed fungi in some detail. He described either the Saffron Milk Cap (*Lactarius deliciosus*) or possibly *Russula alutacea*, and also a fungus with a cap shaped like a priest's head-dress—probably the Shaggy Ink Cap (*Coprinus comatus*).

In the Middle Ages reliance was still being placed on what was really folklore dating back to the times of the Greeks and Romans. This is obvious from reading some of the medieval herbals. In 1601 a book on fungi was published in Italy with clearly recognizable wood-cuts of a number of species. These included edible varieties such as the Morel, Parasol Mushroom and Ear Fungus, as well as the unusual Stinkhorn (*Phallus impudicus*). Soon afterwards the French took the lead with early investigations into the life history of the mushroom.

There are many references to the use of fungi as food in other parts of the world, for instance in central Africa, in India and in the Solomon

The dingy looking Rough-stemmed Boletus (*Boletus scaber*) with its typically scaly stalk is quite good to eat. It is common in woods, especially under birch. Beside it is another common edible fungus—*Cortinarius armillatus*.

Top left
One of the finest-flavoured species is the Chantarelle (*Cantharellus cibarius*). The typical funnel-shaped cap with its lobed margin is clearly shown. The 'gills' are really branched folds which produce pinkish-buff spores. This is an attractively coloured mushroom which looks and smells good to eat.

Bottom left
Horn of Plenty (*Cratarellus cornucopioides*). The dark scaly trumpets of this valued edible fungus are easily missed among the autumn leaves. Basidiospores are produced on gill-like folds on the outside of the funnel-shaped caps.

Right
So called because it may be found on St George's Day, April 23rd, St George's Mushroom (*Tricholoma gambosum*) is a springtime fungus growing in pastures and grassy places, usually on chalky soils and often in rings. It is creamy white with a characteristically thick fleshy stalk.

Islands. In New Guinea species of mushroom (*Agaricus*) and the Jelly or Ear Fungus (*Auricularia*) are still collected and eaten. In Australia the aborigines used to dig up the subterranean sclerotia known as Blackfellow's Bread. This is a species of bracket fungus (*Polyporus mylittae*). Sclerotia of this, and of some tropical fungi, grow to the size of a football and so provide a very considerable source of food. In North America the Yosemite Indians are known to have collected fungi, drying and making them into soup. At the extreme tip of South America, Charles Darwin found a tribe unique in that an Ascomycete fungus (*Cyttaria darwinii*) formed almost the sole item of food in their diet. In the Far East both Japan and China have been collecting, and in some cases growing, species of fungi for food for at least 2,000 years.

In Europe vast quantities of edible fungi were gathered from forest areas. Records show for example, that in Munich in 1900 the market handled 850,000 kg (835 tons) during the year. Half of this enormous quantity was provided by the Cèpe (*Boletus edulis*) and *Boletus scaber*, and the balance was made up of the Field Mushroom (*Agaricus campestris*), the Chanterelle (*Cantharellus cibarius*) and four species of *Russula*. At about the same time, probably as much as 2,000,000 kg (1,964 tons) of wild and cultivated truffles alone (not counting the millions of kilos of other edible fungi) were sold in France yearly. Nothing like this vast quantity of truffles is sold now. At one time on the Continent some towns, including Munich, had special fungus markets and some are still in existence today. Stockholm, Sweden, has a particularly good market and at one time up to three hundred different species could be seen there.

Fungi grow in many different places and at almost all times of the year. In the north temperate zone and in Europe in particular, because fungi like warmth and moisture, June to November are the most prolific months with the main flush from the beginning of September to late October. However, the fruitbodies of some species appear in the spring, for instance the morels (*Morchella*) and the St. George's Mushroom (*Tricholoma gambosum*) which shows in grassy places as early as April and on into July. The dark greyish caps of *Clitocybe rhizophora* appear very early in April and through to May. This species is found in Scandinavia and other parts of northern Europe, but not in Britain. *Hygrophorus hypothejus*, which is fairly rare in Scandinavia but common elsewhere in much of Europe, is strictly winter fruiting, appearing in pinewoods in November and December after the first frosts of the season. The deep yellow-brown caps of the Velvet Foot or Winter Fungus (*Flammulina* or *Collybia velutipes*) may be collected through the winter between September and April. The Oyster Mushroom (*Pleurotus ostreatus*) and the Ear Fungus (*Auricularia auricula*) may be found at any time of the year. All of these species (except possibly *Hygrophorus*) are edible.

Left
The Cheese Cap or Mist Fungus (*Clitocybe nebularis*) is a woodland fungus which grows in clumps or rings and is easily recognized by its large size and delicate pale grey colour. It smells like cottage cheese and is enjoyed by some, but regarded as inedible by others.

Below
Nowadays mushroom cultivation is a controlled process with careful attention to cleanliness, since attacks by insects, bacteria and other fungi can cause serious reduction or total loss of crop. The mushrooms are grown in trays or beds of sterilized compost and are picked daily for two to four months.

Although in small communities it is possible to meet the demand for edible fungi simply by collecting them, larger quantities can be produced only by cultivation. This has been done for hundreds of years in China and Japan. We know from the writings of Dioscorides that the

Above
With its sometimes distorted white or pale yellow caps and blackish scales, *Lentinus tigrinus* has an apt scientific name. It is edible when young and is related to *L. edodes*, the valued 'Shiitake' grown in the Far East.

Greeks and Romans also made some attempts at cultivation. The most widely grown of all species is the cultivated mushroom (*Agaricus bisporus forma albida*) which produces its spores in pairs. The first reference to mushrooms as a crop was printed in France in 1600 and in 1678 a demonstration before the Academy of Sciences in Paris showed how the 'white threads' which develop in the soil under mushrooms, would in turn give rise to more mushrooms. In the 1890s at the Pasteur Institute spores were germinated and a sterilized mushroom spawn produced. This gave the French a virtual monopoly until the 1920s when details were published in the U.S.A. of modern methods of spawn production. The industry then rapidly became established in other parts of Europe, in the U.S.A., South America, Australia and New Zealand and also in parts of the Far East. In 1975–76 the annual weight of cultivated mushrooms produced in Britain was 48,600,000 kg (49,000 tons) and in the same period the world output was nearly 250,000,000 kilos (245,000 tons). The crop is a rewarding one since it is possible to produce up to 81,300 kg (80 tons) per 0·4 hectare (1 acre) in a year.

Mushroom spawn is prepared under laboratory conditions by inoculating barley or other grain. It is then spread over sterilized straw or leaf litter compost in beds under cover in sheds or, as in France, in caves and tunnels. At a temperature of about 24°C (75°F) the mycelium grows freely and is then covered with a thin layer of sterilized soil. Mushrooms appear in six to eight weeks, cropping continuously for two to four months. Fresh compost must be used for each crop, each ton of compost yielding up to 180 kg (400 lb) of mushrooms.

Other cultivated varieties include truffles–the black truffle of France (*Tuber melanosporum*) and the white truffle of northern Italy (*Tuber magnatum*), both of which are grown on a small scale in the south of France. The Padi-straw Mushroom (*Volvariella esculenta*) has been grown on rice straw by the Chinese from early times. Species of *Volvariella* have a very large baggy volva, hence the Latin name. This species is related to the ten or so uncommon European species of *Volvariella*, two or three of which are edible. The Padi-straw Mushroom is grown out of doors and is marketed both fresh and dried. The industry is so successful that this species is now widely grown in south-east Asia, Madagascar and West Africa, and is the third largest mushroom enterprise in terms of world trade. The Chinese also export the Wood or Cloud Ear (*Auricularia polytricha*) which is closely related to the common European and North American Ear Fungus (*Auricularia auricula*). This is a Jelly-fungus with a characteristic gelatinous yet crunchy texture which is extremely popular with the Chinese, both for eating and for medicinal purposes. It resembles *A. auricula* in shape but is more grey in colour when young, with a down of soft hairs on the outside. There was at one time a flourishing export trade from New Zealand to Hong Kong. Nowadays the Chinese collect it, but also cultivate it on oak logs.

Volvariella speciosa is an uncommon edible mushroom which generally requires richly manured ground, straw or compost. Closely related is the Padi-straw Mushroom (*V. esculenta*) which is cultivated by the Chinese on rice straw.

Yet another fungus cultivated in China and also now in Japan is 'Shiitake'. This is *Lentinus edodes*, a gill-bearing fungus growing commonly in tufts on wood, and which is not found outside the Far East. There are, however, a few related species in the north temperate zone. It has a tough yet pliable cap, with cracks giving it a checkerboard appearance, and thin saw-edged gills. Its cultivation has become commercially viable. The process is a lengthy one (taking about two years), but as crops can be gathered over a period of months the results are financially worthwhile. Dried Shiitake needs to be soaked in water for about thirty minutes to soften and can then be used in a variety of meat or fish dishes. There has been a suggestion that Shiitake might be grown in Europe, or that one of the three European species might be substituted, as all are edible when picked young. Such an operation could be carried out in forest or woodland areas, but as the process is rather labour-intensive there are doubts as to its profitability.

Enthusiasm for collecting edible fungi varies from country to country. The British are possibly the least adventurous of all, tending to gather only Field Mushrooms, but on the Continent and eastwards to the U.S.S.R. the picture is quite different. In America too there is often a strong tradition handed down from European forebears, and this is helped by the cosmopolitan nature of the fungi–many common species being found throughout the north temperate zone.

The best way to familiarize oneself with the safe fungi is to collect with someone who is experienced. In addition, there are good handbooks available which are well illustrated in colour. When checking a specimen against the description, points to note include: does it agree with the description in size, shape and colour? Where is it growing? Does the fertile part of the fruitbody have pores, teeth or gills, and of what colour? Is the cap hairy or scaly, or is it smooth, sticky or dry? Does the stalk have any special characteristics? When bruised does the fungus change colour? Is a milky

Below
The Horse Mushroom (*Agaricus arvensis*) is larger than the Cultivated Mushroom and the white or fawn cap may have fragments of the partial veil at the edge and may often bear brownish scales. The ring is large, white and double.

Right
Blue-stalks or Blewits (*Tricholoma personatum (saevum)*) is, like the closely related Wood Blewits (*T. nudum*), a valued edible species growing in open grassland, often in rings.

Bottom right
The general form of the tasty Grisette (*Amanitopsis vaginata*) is similar to *Amanita*, but there is no ring. The long baggy volva is deeply buried in the ground. Both this and the closely related Tawny Grisette are good in stews or omelettes.

juice exuded? (The whole genus *Lactarius* does this, hence the Latin name.) Are the gills covered by a membrane in young specimens, and when this tears does it leave remnants, e.g. as a ring on the stalk? Is the whole fruitbody enclosed in a membrane when young? Are there scattered patches or warts on the cap? Is there a torn sheath (volva) at the base of the stalk? *This is especially important when identifying possibly dangerous Amanitas.*

Always collect the whole specimen–one of the most important features of the poisonous Amanitas is the sheath at the base of the stalk, which may not be noticed if the stalk is broken off short. Most collectors develop an eye for edible species. When in doubt discard the specimen and *always* reject very old or shrivelled specimens which you come across.

Of the edible fungi collected by enthusiasts in many parts of the northern hemisphere some are outstanding for flavour. They include several species of *Boletus* especially the Cèpe (*B. edulis*). This has a brownish cap with yellowish-white pores and a stout brown stalk. It grows in beech woods throughout Europe. The Chantarelle or Girolle (*Cantharellus cibarius*) smells of apricots and has a similar coloured funnel-shaped cap. It is usually found in beech woods. The brownish-black Horn of Plenty (*Cratarellus cornucopioides*) is called by the French 'trumpet of death' and also grows in beech woods in autumn, usually in groups. Naturally growing species of edible mushroom include the Field Mushroom (*Agaricus campestris*), the

The deep cream to tawny caps of the Little Cluster Fungus (*Pholiota (Kuehneromyces) mutabilis*) are produced in very large tufts on the stumps of deciduous trees. They are well worth collecting, especially for adding to stews.

Horse Mushroom (*A. arvensis*) and the Wood Mushroom (*A. silvicola*), as well as the wild type *A. bisporus*. The genus *Agaricus* was formerly known as *Psalliota*.

All species of *Agaricus* have a more or less white cap, smooth or with brownish scales, and with pink to chocolate brown gills. The stalk has a ring but no volva. The fine-flavoured Field and Horse Mushrooms were formerly common but are decreasing in frequency. Both may be found from late summer to autumn growing in pastures and meadows, sometimes in rings. The Horse Mushroom has a large yellowish cap with greyish gills and smells of aniseed. The Wood Mushroom resembles the Horse Mushroom in colour but has a double ring. One of several mushrooms to be avoided as it induces sickness in susceptible people is the not uncommon Yellow Staining Mushroom (*Agaricus xanthoderma*). Both cap and stalk immediately turn a pronounced yellow when bruised, gradually changing to brown. In North America there is a good edible mushroom which also bruises yellow. This is *A. abruptibulba* which grows in rings or groups from July to September, usually among pine needles. The large cap is borne on a stalk with a pronounced bulbous base. These are not the only species of *Agaricus* which change colour. The edible Red-staining Mushroom (*A. langei*), found in deciduous woods and grassy places, turns bright red when bruised.

All species of *Boletus* have small pores beneath the cap which are openings to vertical tubes within which the spores are formed. They are fleshy, sometimes short and

In dry weather the caps of *Pholiota (Kuehneromyces) mutabilis* lighten in colour, but turn a deep rust shade when wet, as shown here.

squat, with often dingy caps and pores which are usually yellow, greenish or brown. This is a large genus with some seventy north temperate species, the great majority being edible although they lack the special flavour of the Cèpe. Curiously, cows have been known to become addicted to eating *Boletus edulis* and indeed another edible species, *B. scaber*, is known in Norway as the Cow Fungus. Among other good boleti are *B. badius* which grows mostly under conifers and has a mid-brown cap and yellowish-green pores, and also *B. cyanescens* which has a hollow stalk bearing a brownish cap with white to yellow pores. In both these species the flesh turns a surprisingly bright blue when bruised. Despite this lurid colour both are good species to eat. *Boletus luteus* grows among grass in coniferous woods. The cap is slimy, brownish-yellow with yellow pores, and the stalk has a distinctive purple-brown ring. Both this and the next species are much valued in Europe for eating. *B. testeoscaber* is found under birch trees and conifers in summer and autumn. It has a reddish-orange downy cap with greyish-brown pores. When cut, the flesh slowly turns dark grey. *Boletus erythropus*, found under conifers, which has a dark brown cap with yellow-green undersurface and red pores, and the somewhat similar *B. luridus* which grows in deciduous woods, were formerly regarded as poisonous. However, they are edible if boiled. When cut, both species change colour from white to blue-green. It is certainly advisable to be able to distinguish the uncommon Devil's Boletus (*B. satanas*), which

Left
The photograph shows a mature specimen of the delicious Shaggy Parasol Mushroom (*Lepiota rhacodes*). The very shaggy cap and stout stalk are typical. The gills are covered when young but the veil breaks, leaving a thick and movable ring.

Above
Wood Blewits (*Tricholoma nudum*) is a fine edible species which grows in rich soil in woods, among leaves and even on compost heaps. All parts are lilac, but the cap later turns reddish-brown.

although not deadly is definitely poisonous. It grows on chalky soils in deciduous woodland and has a grey-green cap with blood-red pores and turns blue when cut.

The Parasol Mushrooms, *Lepiota procera* and *L. rhacodes*, are large handsome cap-and-stalk fungi with a brown, conspicuously scaly cap and long stalk with a large double ring. The cap is easily removed from the stalk and when fully expanded has a distinct umbrella or parasol shape. The ring can be moved up and down the stalk further simulating a parasol. *Lepiota procera* has a cap 10–20 cm (about 4–8 in.) across and a scaly stalk up to 28 cm (11 in.) high. *Lepiota rhacodes* is shorter, more squat in appearance, with a smooth stalk and smaller cap. Both parasols are found in wood clearings, gardens and open grassy spaces between July and November, and *L. procera* is often found on cliff tops by the sea. Both need to be picked young as they tend to become leathery. They dry well, or can be cooked fresh as a vegetable with a white sauce, or stuffed with any savoury mixture and baked. *L. procera* is said to taste like almonds or Brazil nuts.

The attractively coloured Blewits were at one time collected and sold. Wood Blewits (*Tricholoma nudum*), also known as the Great Violet Rider, appears in woodlands from autumn into winter. It is lilac rather than blue, with lilac gills, and is known to contain minute amounts of a substance which attacks red blood cells, but the poison is destroyed on cooking. Blewits (*T. personatum*) has a slightly larger cap with gills which are white or flesh-coloured. It grows from autumn into winter, usually in rings, on pastures and downs. Another esteemed species is the cream or buff St. George's Mushroom (*T. gambosum*). This grows in grassy places especially on chalk or limestone soils but is never common. It often forms fairy rings and is essentially a springtime fungus. It should be collected in dry weather and cooked carefully to avoid toughness. It

The striking greyish-blue cap and shell-like shape of the Oyster Mushroom (*Pleurotus ostreatus*) is well portrayed here. It usually grows on living trees, especially, as depicted, on beech.

should not be confused with *Entoloma lividus* (also known as *Rhodophyllus sinuatus*) which is poisonous, although not deadly, and which can be distinguished by yellowish gills and salmon-pink spores. The Yellow Knight Fungus (*T. flavovirens* or *equestre*), much esteemed on the Continent, grows in sandy pinewoods, appearing in August and September, with a yellowish-green cap and sulphur yellow gills. In Japan a species of *Tricholoma* grows under pine trees and is collected on a commercial basis since it has a particularly fine flavour. This is 'Matsutake' (*T. matsutake*), and gourmets put it on a par with the Cèpe (*Boletus edulis*). It has not yet been cultivated.

In France the delicate-flavoured Oyster Mushroom (*Pleurotus ostreatus*) is sold in the markets. Although a cap-and-stalk gill-bearing fungus, it rather resembles a bracket fungus in shape and appearance, growing in colonies on dying or dead trunks of beech and other trees. In parts of eastern Europe it grows on logs for much of the year; it could undoubtedly be cultivated. Its common name refers to its shape and greyish-blue colour. Only young specimens should be used for eating. Also valued is the 'Bouton-de-guetre' (gaiter button) or Champignon, known in Britain as the Fairy Ring Mushroom (*Marasmius oreades*), which may be found on garden lawns and other

The Blusher (*Amanita rubescens*) is common in woodland and is one of the few edible amanitas. Bruising causes a characteristic red stain to develop.

stretches of grass. Its flesh smells faintly of sawdust, but has a pleasant flavour when fried or made into an omelette. It also dries well.

Despite its name (*Lactarius deliciosus*) the Saffron Milk Cap is not rated highly. It should be washed thoroughly before being grilled and served with a knob of butter. Possibly it has been confused with the finer-tasting *L. sanguifluus* of southern Europe. Nevertheless *L. deliciosus* is regularly gathered in France, where it is called 'vegetable sheep's kidney', and where it is sometimes used in place of the more expensive truffle. It is found under conifers between August and November and has an orange cap with deeper concentric zones. When cut it exudes a deep orange milk which turns green in the air.

The usually elongate caps of the ink caps (*Coprinus* species) turn black at the base because the gills undergo a process of auto-digestion, but if picked before this happens the scaly caps of the Shaggy Ink Cap (*Coprinus comatus*) are extremely tasty when cooked. They are probably best steamed or made into a filling dish with bread-crumbs and grated cheese. The Shaggy Ink Cap (also called Lawyer's Wig) is often found in clusters on rich soil by roadsides, in fields or on rubbish tips. Appearing from May until November it is probably the best-known species of *Coprinus*.

Despite their extremely bright colours several species of *Russula* are good to eat. The genus is characterized by its rather short squat shape, and medium to large caps with brittle white to yellow gills. All are woodland species. The Sickener (*R. emetica*), with its scarlet cap and white gills, is aptly named and should be avoided. The best species for eating are *R. claroflava* with bright yellow cap and primrose-coloured gills, *R. vesca* often found under oaks from August to November, which has a pinkish-buff to dull red cap, and *R. cyanoxantha* which has a cap varying from violet to greenish-blue.

The Plum Agaric (*Clitopilus prunulus*), also called the Miller from its mealy smell and flavour, is a common and highly edible fungus of woods and pastures, appearing between June and November. It is easily recognized by its whitish, somewhat funnel-shaped cap with a wavy margin and with the gills running part way down the stalk.

Pholiota (Kuehneromyces) mutabilis, sometimes called the Cluster Fungus, appearing in deciduous woods (rarely with conifers) between April and December, may have as many as 400 fruitbodies tightly packed in tufts. In dry weather the cap is dark cream, but when wet turns a rich rust colour. It is borne on a characteristically curved slender stalk which has a conspicuous ring. It is a common fungus well worth picking. *P. aegerita* which grows on elm and poplar was especially favoured by the Greeks and Romans. The common Bootlace Fungus or Honey Fungus (*Armillaria mellea*) is quite good to eat despite its unpleasant smell and bitter taste—both of which disappear with cooking, especially if first blanched in boiling water. It must be picked young and the tough stems discarded, but even then some find it too rich and indigestible.

Although the amanitas are notorious for their extremely poisonous varieties they include the highly tasty grisettes *Amanita (Aminotopsis) vaginata* and *A. fulva*, the red-staining Blusher *A. rubescens*, and the delicate-flavoured *A. caesarea* which is not found in Britain. The grisettes (*A. vaginata* and *A. fulva*) are sometimes put in the separate genus *Aminotopsis* because they lack a ring. *Amanita vaginata* has a grey cap and stalk with a torn baggy greyish volva at its base. It is found in several varieties including one white and one dark grey. *Amanita fulva* has a cap orange to mid-brown in colour, with the stalk and volva somewhat paler. Both grisettes are woodland fungi, but *A. vaginata* favours beechwoods whereas *A. fulva* prefers birches. The Blusher (*A. rubescens*) stains red on handling. It has a reddish-brown or greyish-brown cap with off-white warts. The stalk bears several rows of warty patches which are all that remains of the volva. It grows in both deciduous and coniferous woods, also on heaths, in summer to late autumn. It is common and widespread in north temperate regions. By contrast, the handsome *A. caesarea* is found in North America and southern Europe (including the Mediterranean countries), but is absent from the more northerly parts of Europe including Britain. It is not unlike the poisonous Fly Agaric (*A. muscaria*) but has noticeably yellow gills and stalk and lacks the white patches on the cap. There are about another six edible species which are found in North America

Left
Paxillus involutus is one of the commonest woodland fungi, with a yellowish-brown flat or slightly funnel-shaped cap, typically with a rolled margin. It is edible if blanched before cooking.

Above
Well-named, the Ugly Milk Cap (*Lactarius turpis*) is a common fungus with a thick slimy cap up to 15 cm (nearly 6 in.) across. When cut, it exudes an acrid white milk yet despite its unprepossessing appearance it is edible.

Right
The good flavour of *Russula vesca* makes it well worth collecting. Its pinkish-fawn white-rimmed caps are found in deciduous woods, especially under oaks.

Amethyst Deceiver (*Laccaria amethystina*). This attractive fungus has widely spaced dusty looking gills. Despite sometimes smelling of garlic it is quite good to eat.

and parts of Europe, but they are not common.

The genus *Clitocybe* is characterized by funnel-shaped caps with white-spored gills running part-way down the stalk. The better-tasting species include the Cheese Cap or Mist Fungus (*C. nebularis*), the flesh-coloured *C. geotropa*, and the cream-coloured *C. gigantea* (also known as *Leucopaxillus gigantea*). *Clitocybe nebularis* is a soft pale grey throughout and has a smell like cottage cheese. Many praise it but some find it peppery and indigestible. *C. geotropa* is larger, with a cap up to 20 cm (8 in.) across. This too is common, often growing in large rings, and appearing between September and November. The largest edible species of *Clitocybe* is *C. gigantea* with a truly funnel-shaped cap up to 30 cm (about 1 ft) across. Although not common it can be found growing in late summer and autumn in various grassy places including roadsides and hedgerows. It often forms rings, and because of its size may kill the grass. On the Continent it is extensively eaten, although mildly poisonous when raw. One species of *Clitocybe* fairly common on the Continent and in Scandinavia, although not found in Britain, is the white-capped *C. conglobata*. This grows in enormous tufts of hundreds of specimens. It has a good flavour and is well worth collecting. Great care should be taken not to pick *Clitocybe rivulosa* or *C. dealbata* by mistake. These are both common in grassy places (like the edible varieties), but are smaller–the cap rarely exceeding 4 cm ($1\frac{1}{2}$ in.) across. Both species are highly poisonous and sometimes deadly. *C. rivulosa* could be picked together with the Fairy Ring Mushroom (*Marasmius oreades*) which is much the same size but rather darker.

Most edible fungi can be cooked like mushrooms but some are improved by first boiling, either to soften them or to remove unpleasant flavours. Two species improved in this way are the Wood Hedgehog (*Hydnum repandum*), and the Beefsteak or Ox-tongue Fungus (*Fistulina hepatica*) which is a bracket fungus. They can then be cooked in butter or added to stews. The Wood Hedgehog–known in Italy as Stercherino (little hedgehog)–has teeth on the underside of its pinkish-buff cap. It grows in groups in deciduous woods between August and November. The Beefsteak Fungus is very short-lived for a bracket fungus, not usually lasting longer than three weeks. It is common on the trunks of oak or sometimes sweet chestnut and is a parasite causing a brown rot of the wood. It appears between August and November growing up to 40 cm (about 1 ft 4 in.) across. Appropriately named it resembles a piece of raw meat or liver, with reddish flesh which exudes a red juice when cut. The pores underneath are pale yellow.

Two other species of bracket fungi which are edible when young are both species of *Polyporus. Polyporus frondosus* (also known as *Grifola*) has the attractive name of Hen (or Chicken) of the Woods, its clustered greyish-brown appearance resembling a tousled bunch of feathers or a mother hen sheltering her chicks. Widespread throughout the northern temperate zone, it appears in summer and autumn, usually on oak trees. Despite a smell of house mice and a somewhat leathery texture, it is good pickled in spiced vinegar for adding to savoury dishes. The bright yellow Sulphur Shelf Mushroom (*Polyporus sulphureus*), also known as *Polypilus* or *Grifola sulphureus*, is found in Europe and North America, but the two regions have markedly different views on its edibility. In Europe it is usually dismissed as 'edible but worthless'. In America it is considered to be one of the four best and safest mushrooms, the others being the morels, the puff balls and the Shaggy Ink Cap (*Coprinus comatus*).

Top
The Yellow Smooth-edged Russula (*Russula ochroleuca*) is the commonest of the yellow-capped russulas. It is edible and may be found in all types of woodland.

Above
Chicken of the Woods (*Grifola (Polyporus) frondosa*). This feather-like fungus is sometimes collected for eating. It grows on deciduous trees, especially oak, but is never common.

Puff balls are also Basidiomycetes and all are safe, but they must be eaten young, before the spores are properly formed. The best is the Giant Puff Ball (*Lycoperdon* or *Calvatia giganteum*) known to the French as Dead Man's Head. Its rounded body can exceed 30 cm (1 ft) in diameter. It grows in pastures, orchards and woodlands, sometimes in rings, appearing in late summer and early autumn. It should still be white inside when picked and can be sliced, dipped in egg and breadcrumbs and fried in butter. It is also good if stuffed with sage and onion and cooked covered with slices of fat bacon. Smaller puff ball species are stewed or fried in butter.

The Common Earth Ball (*Scleroderma aurantium*), which superficially resembles a small puff ball, is widespread between July and January. In eastern Europe it is made up into sausages as a cheap substitute for truffles, but is not eaten in Britain, probably because if consumed in any quantity it causes sickness or even unconsciousness.

The Ear Fungus (*Auricularia auricula*) is less often picked yet is quite good to eat, having a firm gelatinous texture. The translucent fruitbody is extremely variable and ranges from flesh-pink to almost black in old specimens. When moist it is gelatinous but becomes tough as it dries. Usually growing in groups at any time of the year, it is common, and in Europe is almost always found on elder bushes, but in North America appears on other trees including oak, beech and elm. The Chinese rate their species *Auricularia polytricha* highly and cook small quantities of the fungus with onions and pig's liver in a thick sweet sherry sauce. *Auricularia auricula* would make an equally savoury dish.

Apart from truffles and morels, relatively few Ascomycetes are edible, chiefly because most species do not produce sizeable or fleshy fruitbodies. Some are acrid or bitter to the taste and a few are highly poisonous, for instance the Lorchel (*Gyromitra esculenta*) which, despite its descriptive specific name, has a doubtful reputation. Truffles are probably the most highly prized of all edible fungi, having a delicate scent and flavour. They grow underground in soil or in leaf-mould at depths from 10 cm (4 in.) to as much as 30 cm (1 ft), and can attain a weight of nearly 1 kg (about 2 lb). They tend to be found in chalky soils with beech or evergreen oak trees. Truffles are hard to trace, but when ripe they exude a strong characteristic odour. This can be detected at a distance of over a metre by a number of animals including pigs, dogs, goats, squirrels, deer and bears. For centuries pigs and dogs have been used for this purpose. Sometimes it is possible to ascertain the position of truffles underground by noticing swarms of the yellowish 'truffle flies' or by noting bulging and cracking of the soil around the base of trees. The latter method is used by Arabs in Mediterranean countries to find another variety of truffle—*Terfezia*. This is sold in markets throughout North Africa and eastwards to Asia Minor.

Left
A common and easily recognized fungus, the Common Earth Ball (*Scleroderma aurantium*) might be mistaken for a young puff ball but is firmer in texture and never develops an apical pore. It is not recommended for eating.

Top right
The spines or teeth on the underside of the convoluted cap of the Wood Hedgehog *(Hydnum repandum)* give this its common name. If boiled before being cooked it is good to eat.

Bottom right
Growing characteristically in beech leaf litter the pleasing colour of the edible Lilac Curtain Fungus or Lilac Thickfoot (*Cortinarius albo-violaceus*) makes it easy to find. The thick stalk which gives its common name is clearly shown.

Below
The photograph shows a typical specimen of the easily recognized Beefsteak or Ox-tongue Fungus (*Fistulina hepatica*). The sticky glistening upper surface, the blood-red colour and tongue-like shape are all highly characteristic of this bracket. Despite the common name of 'beefsteak', it is not particularly tasty.

Left
With their keen sense of smell, dogs can be trained to scent out buried truffles. Poodles are commonly used, especially in France, but mongrels can also be successfully trained.

Right
A striking fungus of open grassland with red cap and stalk, and yellow to red waxy gills is the Carmine Wax-gill (*Hygrophorus coccineus*). This conspicuous and common fungus is edible, but being small is not often collected.

Right
The conspicuous edible Cauliflower Fungus (*Sparassis crispa*) may grow as much as 30 cm (1 ft) across and resembles a large sponge or cauliflower. It is not common, but may sometimes be found on pine stumps.

Below
Not one of the finest flavoured truffles, Summer or Cook's Truffle (*Tuber aestivum*) is nevertheless worth eating. It has a purplish-brown, almost black warty skin and the flesh inside is violet, veined with white.

The best truffles are the two species of *Tuber, T. magnatum* (the White or Piedmont Truffle of northern Italy) and the Black or Perigord Truffle *T. melanosporum* which is still common in parts of France. Neither is found in Britain or North America. The Cook's Truffle (*Tuber aestivum*) is found in beechwood areas of Britain and other parts of Europe. At one time in south-west England the so-called 'red truffle', *Melanogaster variegatus,* which resembles a small reddish-brown potato, was collected for market. It is totally unrelated to the true truffles and is a Basidiomycete somewhat like the earth balls. Truffles have always been expensive and recipes use them in small quantities, with eggs, chicken or cheese. Piedmont truffles are not usually cooked as heat spoils the flavour–they are either used raw or cut into thin shavings and gently warmed in a little butter.

Like truffles, morels are highly

valued Ascomycetes. All are edible (the best being *Morchella esculenta*) and appear in springtime. They seem to require rich soil but sometimes grow on lime-rich sand dunes. They also favour the sites of recent fires, and in Europe at one time poor people burned small areas of woodland to encourage the growth of morels. In France after World War I morels flourished in abandoned trenches and burnt-out house sites. Morels have an unusual appearance, with a short stout yellow stalk bearing a rounded, highly convoluted brownish-grey head. They are generally sliced or split before being thoroughly washed in salt water. They can be fried or cooked with cream and added to egg, fish, chicken or veal dishes, or baked stuffed with seasoned minced meat.

One of the largest edible Basidiomycetes is the white waxy Cauliflower Fungus (*Sparassis crispa*) which grows at the base of pine stumps between August and November, producing a densely branched white head 30 cm (1 ft) or more across. It should be eaten young, either baked with butter and seasonings in a little stock or white sauce, or sliced and dipped in egg and fried. It can also be dried. *Ramaria flava* and *R. botrytis* are similarly densely branched Basidiomycetes with a much more twiggy appearance. They are edible, but unfortunately the darker-coloured *Ramaria formosa*, which tastes very bitter when cooked and is poisonous, can be mistaken for them.

Because of their subtle flavour all freshly picked fungi should be washed as little as possible. Unless the skin of the cap seems tough it need not be peeled, but it is often better to remove the stalks. Once gathered, fungi should be cooked and eaten within six hours as they soon go mouldy. Many varieties can be successfully dried for storage and indeed it is possible to buy cultivated mushrooms, Cèpes and some of the Oriental species dried and packaged, or in tins. Drying is simple. The caps are threaded on thin canes or string, then dried in the sun or in a slightly warm oven with the door open, or over a radiator, the temperature not exceeding 55°C (131°F). In many countries it is usual to boil fungi in several changes of water—a method which is said to make acrid, or even poisonous fungi fit to eat. This is probably true for many doubtful species but certainly not for the dangerous Amanitas.

Recipes for mushrooms of all kinds abound. Excellent mushroom sauces may be found on the Continent. Many consist of a white sauce to which is added cream, eggs, mushrooms and sometimes other ingredients such as chopped walnuts. Stuffings are made by adding breadcrumbs, egg yolks or chopped olives. The French make a simple but tasty supper dish by frying Cèpes, placing them on buttered toast with cheese on top and then grilling.

Left
Although the Fairy Club (*Clavaria pistillaris*) is never found in any quantity it is nevertheless edible and may sometimes be seen in deciduous woods. It varies in size from 10–30 cm (4–12 in.) high. The related branched and fleshy fairy clubs *Ramaria botrytis* and *R. flava* are also worth gathering.

Right
Green Collared Trout. A recipe for this tasty and attractively garnished dish appears in the text.

Mushroom soups include versions made with game such as hare, or with mussels. Mushrooms combine particularly well with shellfish or prawns, and can be mixed with fish to stuff pancakes. A rather unusual combination is that of trout and mushrooms.

Green Collared Trout

Quantities for 4 persons
4 trout
4 open mushrooms
575 g (1¼ lb) cup mushrooms
1 wineglass white wine
2 medium onions
1 small green pepper
1 lemon
salt and pepper

Put the fish and four whole mushrooms into a shallow ovenproof dish. Pour over the wine and season with salt and pepper. Cover and poach in oven 190°C (375°F), Gas Mark 5, for about 25 minutes. Mince mushrooms, onion and half the green pepper. Strain wine from the fish into a shallow pan; add mushrooms, onion and green pepper. Season with salt, pepper and lemon juice and cook over a moderate heat until all the liquid has evaporated. Skin the fish (if preferred). Put the mushroom mixture on to a hot serving dish. Arrange the fish on top. Put a collar of green pepper on each fish and garnish the fish with slices of lemon and whole mushrooms.

Large mushroom caps can be stuffed and baked. Suitable fillings include chopped hardboiled eggs, ham, sausage meat, crab or anchovy fillets. Similar mixtures can be combined with mushrooms to make savoury flans (quiches) or added to potatoes and fried as croquettes. Mushroom caps may also be dipped in batter and deep-fried or may be put in alternating layers with vine leaves, sprinkled with olive oil and slowly baked.

Savoury Ham

Quantities for 4–6 persons
1 medium-sized onion
75 g (3 oz) butter
175–225 g (6–8 oz) mushrooms
350–450 g (12–16 oz) chopped cooked ham
3 dl (½ pint) well-seasoned thick white sauce
25 g (1 oz) breadcrumbs
50 g (2 oz) grated Cheddar cheese

Chop and cook the onion in a shallow pan in 50 g (2 oz) of the butter until soft and golden in colour. Then add the mushrooms, quartered. When cooked, add the chopped ham and stir in the white sauce. Turn the mixture into a casserole, cover with the breadcrumbs and grated cheese and dot with small pieces of the remainder of the butter. Cook in a very hot oven, 230°C (450°F), Gas Mark 8, for about 10 minutes until the top is golden-brown. This recipe is also a good way of using up cold left-over meat.

In Italy liver is often cooked with a mushroom sauce and this is excellent served with either rice or potatoes together with green vegetables. In the U.S.A. mushrooms are used to stuff peppers and, as both are normally obtainable all the year round, this makes a convenient savoury.

A good basic recipe is to place any kind of mushroom in a casserole with butter or oil. Season, cover and cook in a moderate oven–180°C (350°F) or Gas Mark 4–until tender. Serve with a squeeze of lemon juice. The butter may be replaced by soured cream–many Russian recipes use soured cream in this way–or by a little meat stock. Herbs or chopped onions could also be added or a mixture of egg yolks and cream, with or without breadcrumbs.

Birch Polypore or Razor-strop Fungus (*Polyporus (Piptoporus) betulinus*). Although not suitable for human consumption this bracket fungus may form an important part of the diet of reindeer in Finland and Scandinavia.

Mushrooms, as well as truffles, can be used raw. The small button cultivated mushrooms are sliced thinly and added to mixed green salads. The Giant Puff Ball can be similarly treated.

Mushrooms consist of about 90 per cent water so their value is as a flavouring. The calorie values for fungi are lower than those for fish and meat, but similar to those of most vegetables. However, their protein values are higher, the cultivated mushroom for example containing nearly 4 per cent. Unfortunately this is not always easily digested; nevertheless comparing dried vegetables with dried mushrooms, vegetables are far poorer in protein.

Yeast has long been known as an excellent source of most vitamins, especially the vitamin B complex, and most of the B group are also present in the larger fungi, vitamin B_1 being more frequently present than B_2. Vitamin C is never present in any useful amount but vitamin D occurs in at least four common mushrooms: the Cèpe, morels, the Cultivated Mushroom and the Chanterelle, which is also quite rich in vitamin A. Mushrooms are relatively rich in copper and iron.

When investigating the food value of fungi, laboratory rats were given mushrooms as a source of protein and then compared with a group fed on cheese. Mushroom-fed rats put on 30 per cent more weight. Fungi have also been successfully fed to pigs, poultry and even fish in place of the more usual protein supplements. Cattle in Finland are commonly fed on fresh or salted mushrooms. In Scandinavia and Lapland reindeer eat lichens, particularly Reindeer Moss (*Cladonia rangiferina*) and will also feed on young bracket fungi such as *Polyporus betulinus*. The importance of lichens and fungi to reindeer is evident when one realizes that 20 per cent of the meat produced from these animals can be attributed to the fungal and lichen part of their diet. In the wild a surprisingly wide range of animals eat mushrooms and other fungi. Squirrels consume a number of different species including truffles and Cèpes. They also eat the Fly Agaric without ill-effect. Other rodents, wild deer and game birds regularly eat mushrooms when available. Even more surprising is that carnivorous species such as bears and wolves are known to feed on mushrooms.

In times of adversity people in many different parts of the world have subsisted on a diet consisting almost solely of edible fungi. With a growing gap between world population and available food supplies it may well be that the fungi, including food yeasts, will come to be more generally valued.

Poisonous fungi

The Death Cap (*Amanita phalloides*) is aptly named since it is one of the deadliest of all fungi. The cap colour may be paler than that shown but the distinguishing volva and ring are very clear.

Fungi have probably been eaten from prehistoric times since they constitute a relatively abundant source of food. Before early man cultivated crops, he collected plants and hunted animals and it is certain that Palaeolithic man would have collected fungi. By a gradual process of trial and error, species with a repulsive flavour or smell, and others producing burning sensations in the mouth, would have been avoided. Thus only those fungi with a pleasant or mild taste would have been used, although even this would not have been an absolute safeguard since the most deadly of all—the Death Cap (*Amanita phalloides*)—has practically no smell or taste when young. Doubtless in the course of time some species which are unpleasant or poisonous when raw would have been cooked and then eaten. For example the Maoris long ago discovered that some fungi, inedible when raw, could be made fit to eat by wrapping them in leaves and then burying them for some time in hot ashes. It is possible that Stone Age man might have similarly treated certain fungi.

The earliest known classical mention of fungi is in the writings of Euripides (480–406 BC) and Hippocrates (460–*c.* 377 BC), and it is interesting that both writers refer to cases of mushroom poisoning. The Romans were fond of a number of different kinds of fungi and, in particular, the 'egg mushroom' (*Amanita caesarea*). This still grows in abundance in Mediterranean countries and as far north as Sweden, although is not found in Britain. Possibly the most famous person poisoned by fungi was the Roman Emperor Claudius I (10 BC–54 AD). He was served by his fourth wife, Agrippina, with a dish of mushrooms which are thought to have been adulterated with the Destroying Angel (*Amanita virosa*) which is virtually as deadly as *Amanita phalloides*. Often, to prevent accidental or deliberate mishaps, slaves were used to taste foods, but presumably the Emperor's wife was regarded as above suspicion.

A number of popular tests for poisonous fungi are known, and some of these have been handed down from classical times. It has been said that poisonous species will not peel, or will blacken silver coins or spoons when cooking, or that they will coagulate milk or turn a cut onion bluish or brown. None of these statements necessarily holds good, and fungi which do not react are not always safe to eat. It is also claimed that fungi which have been nibbled by slugs, rabbits or squirrels are safe for man but the evidence for this is not reliable. The poisonous effects of some fungi may be due to a variety of causes. For example, it may be a case of simple allergy, just as some people are unable to eat eggs or strawberries without ill effects. Fungi are undoubtedly rather indigestible and can cause stomach upsets if eaten in excess or without accompanying food such as bread or meat. Unless they are dried, fungi should always be eaten freshly gathered, as partly decomposed fungi can produce symptoms similar to those of ptomaine poisoning. (Ptomaine is a substance produced during putrefaction of animal or plant proteins.)

Some fungi are certainly extremely poisonous but the amount of poison present varies. The poison itself may subsequently be modified by heating, boiling or drying.

The annual death-rate from fungal poisoning varies from country to country. Britain has on average one or two fatalities a year, the U.S.A. about fifty but, because of the greater amount of fungi eaten, there are many more deaths on the Continent. For example, in 1948 in Germany, as many as two hundred deaths were recorded. Various attempts have been made to reduce the numbers of such cases. Thus in the U.S.S.R. batches of mushrooms were, and presumably still are, inspected for poisonous species, and France at one time had 'mushroom inspectors' attached to local markets.

Nowadays more species are coming under suspicion. In Britain a Ministry of Agriculture, Fisheries and Food Bulletin names only nine suspect species whereas the leading authority John Ramsbottom, formerly of the British Natural History Museum, lists twenty-five species – about one or two per cent of the larger fungi. In the eastern U.S.A. Buck names fifty-three poisonous species and most of these also occur in Britain. On a worldwide basis probably about three hundred species should be regarded with suspicion. Most cases of poisoning are due to species of *Amanita* and it is therefore extremely important for collectors to be able to recognize this genus. Amanitas are widespread in distribution and few of the numerous species are edible. Consequently it is far safer, unless one is knowledgeable, to leave this genus completely alone. Amanitas are fairly large cap-and-stalk gill-bearing fungi with a large basal volva or sheath, a bulbous stalk with a ring, and usually warts on the cap. The gills are white, producing white spores, and hang free from the lower surface of the cap. The volva may not be noticed if part of the stalk is left below ground and therefore collectors should always be careful to pick specimens completely whole.

The Death Cap (*Amanita phalloides*) is fairly common throughout Europe, appearing in deciduous (sometimes coniferous) woods from August until October. The cap is at first rounded, later becoming flat-

Left
The elegant Destroying Angel (*Amanita virosa*) is one of several pure white amanitas, all of which are extremely poisonous. The slender form, conspicuous volva and ring are all very characteristic of this species.

Above
Inocybe fastigiata is a rather small fungus which is not infrequent in deciduous woods, but should be avoided since it is known to be deadly poisonous.

tened, and grows about 10 cm (4 in.) across on a stalk up to 20 cm (8 in.) high. Its colour is of varying shades of grey to olive-green, fading to nearly white. The volva has jagged edges which stand well clear of the stalk, and the ring this bears hangs down loosely. Fresh specimens have no smell but later become characteristically foetid and sickly. If eaten by mistake the effect is devastating, and almost always fatal if more than one average-sized cap is consumed. The main symptoms rarely appear before 12 hours and are those of violent gastroenteritis. Within 2–3 days the patient passes into a deep coma and usually dies after 5–10 days. The Death Cap is a common European species, especially in France and the Channel Isles. Until the early 1970s it was thought to be rare or absent from the U.S.A., but now appears to be established, certainly in the eastern states. It is believed to have come from Europe as pieces of mycelium attached to the root systems of imported trees. Much more common in the U.S.A. is *Amanita brunnescens*. This has a more fragile volva and bruises easily on handling, developing brown stains on the stalk. This too is poisonous but its toxins differ from those of other Amanitas. The Death Cap is by far the most deadly of all fungi and is responsible for over ninety per cent of all human deaths from fungal poisoning.

The toxins present in the Death Cap have been shown to be much more complex than previously thought. One group (formerly known as phalloidin) comprises six compounds known as phallotoxins; another much more toxic group (formerly called amanitin) comprises five compounds known as amatoxins—which are much the more toxic; and there is also a recently discovered group called myrio-amanins. Amanitin attacks liver and kidneys, normally causing irreversible damage to these organs, but the finding of an antidote—thioctic acid—during the past few years holds out some hope for Death Cap victims, and it has already been used with some success. In 1973 yet another treatment was tried out on a victim who was suffering from severe liver failure; the blood system was linked to a dialysis machine and the patient subsequently made a complete recovery.

Because of its distinctive appearance and frequent occurrence, the Fly Agaric (*Amanita muscaria*) is perhaps the most widely known of all cap-and-stalk fungi. It develops the familiar white-spotted scarlet cap and grows, generally in clusters, in pine and birch woods between August and November. One might think that its dramatic colouring would present a warning signal but it may prove attractive to children. Indeed, in some parts of the world it is eaten in very small quantities as a stimulant, yet if consumed in larger amounts it causes violent stomach upsets and induces delirium and coma. The effects are serious but rarely fatal since the poisonous constituents (muscimol and muscarine) do not include any of the deadly Amanita toxins. Nowadays we have a successful antidote to muscarine; this is atropine, which was formerly obtained from the flowering plant deadly nightshade and which is now often synthesized. The common name, Fly Agaric, derives from its use in a number of countries to kill flies. Portions of the cap are broken up in milk and the resultant liquid is both attractive and fatal to flies since the fungus, especially the skin of the cap, contains a highly effective organic insecticide.

Left
All inocybes should be regarded with suspicion as many are difficult to identify and a number are extremely dangerous if eaten. This species is *Inocybe calospora*.

Right
Usually growing in woodland, and often preferring rich soil, *Inocybe maculata* is a poisonous fungus which has a misleadingly pleasant smell.

Below
One of the smaller highly poisonous species of *Clitocybe*, *Clitocybe dealbata* grows in grassy places but may also be a pest in mushroom houses.

Another highly poisonous species of *Amanita* is the Panther Cap (*A. pantherina*). This produces severe effects on the nervous system such as visual disturbance, delirium and muscle spasms, but is only rarely fatal. It contains some of the toxins found in the Fly Agaric and in Japan is similarly used to kill flies. Unfortunately it resembles an edible species–the Blusher (*Amanita rubescens*)–which changes to a reddish colour when bruised. Both have a greyish-brown cap with conspicuous white warts, but in the Panther Cap (sometimes called the False Blusher), the gills remain white and the flesh never changes colour.

Amanita virosa has already been mentioned in connection with the death of Claudius I. Its common name of Destroying Angel derives from its pure white colour. Its virulent effect is due to the presence of toxins similar to those of the Death Cap. It is a rather handsome fungus, white throughout, and with a scaly stipe. There is a ragged volva at the base, but the ring is frequently torn or absent. The cap is 5–9 cm (2–3½ in.) across, conical when young, with a sticky, rather shiny appearance. The Destroying Angel grows in deciduous woods, usually on poor soils, and appears between August and October. It is found throughout Europe and the U.S.A. as is another pure white poisonous species–*A. verna*–which, despite its name, appears from May right through to October. *Amanita bisporigera* (with spores produced in pairs instead of in fours) and *A. tenuifolia* are also white and are common in the U.S.A., but the latter is restricted to the southern states.

Another poisonous amanita is the Italian fungus *A. aureola*. This has an orange-red cap with white gills, rather like that of the Fly Agaric (which is also found in Italy), but without the white warts. *Amanita citrina (mappa)* with its lemon-yellow cap smells of new potatoes when it is cut. Formerly it was regarded as poisonous and it certainly tastes unpleasant although is not unsafe to eat.

Rather surprisingly the same poisonous compounds may occur in unrelated fungi. Thus muscarine found in the Fly Agaric is also present in some species of *Clitocybe* and *Inocybe*, both of which also include a number of harmless species. Both are widespread in Europe and the U.S.A. *Clitocybe* has several large good-tasting species, for instance *C. geotropa* and *C. gigantea* which, like all clitocybes, have concave, somewhat funnel-shaped caps producing white spores. However, several small whitish clitocybes are common and poisonous, sometimes deadly. Such species are distributed throughout Europe and also in the U.S.A. The False Champignon (*Clitocybe rivulosa*) grows in tufts or rings in short grass and is often associated with the edible Fairy Ring Mushroom (*Marasmius oreades*). This is unfortunate since if the two are cooked together by mistake the results can be severe. *Clitocybe dealbata*, also growing in grass, is equally small and insignificant and equally poisonous. Both species are whitish-fawn in colour with pale gills running part way down the stalk and both appear in late summer and autumn.

It is unlikely that anyone would deliberately pick *Inocybe* to eat, as the species are rather small. All are woodland fungi with brownish or pinkish caps 2–5 cm (¾–2 in.) across and producing cinnamon-brown spores. Often the cap has a central hump and an earthy, sometimes fruity smell. The Red-staining Inocybe (*I. patouillardii*) is probably the most dangerous European species with several fatalities to its credit. Fortunately it is not common. It is found on chalky soils often in beech woods, and turns bright pink when bruised or broken. The Earthy Inocybe (*I. geophylla*) is white or lilac in colour and like *I. fastigiata* is common in Europe and also in the U.S.A. This latter species is the largest of the dangerous species of *Inocybe* with a yellowish-brown cap about 10 cm across. If any are accidentally included with a dish of cooked mushrooms the victim quickly feels sick and dizzy, the heart slows and breathing becomes difficult. Although serious, the effects are rarely fatal.

Most species of *Panaeolus* have a distinctly bell-shaped cap and grow on animal droppings or on rich garden soil. *Panaeolus semiovatus*, here, differs from most of them in having a distinct ring.

Left
Also poisonous, *Clitocybe rivulosa* is found in short grass, sometimes in rings as seen here. It may grow together with the Fairy Ring Mushroom (*Marasmius oreades*) and could then be picked, with disastrous results.

Below
The clear scarlet cap and white gills of The Sickener (*Russula emetica*) make this a conspicuous woodland fungus. If eaten it lives up to its common name. Nearby is a 'twiggy' lichen.

One does not associate the genus *Agaricus* with poisonous fungi, but in fact it is safer to regard at least two species with suspicion. Of these the most common is the Yellow-staining Mushroom or Yellow Stainer (*A. xanthoderma*). The Yellow Stainer looks very much like the Field Mushroom and grows in similar places, but can be distinguished by touching it. Immediately a characteristic bright yellow stain appears. Many people can cook and eat the Yellow Stainer without any ill-effects, but others (about ten per cent) seem much more sensitive, reacting with violent stomach upsets which can take about three days to clear. In very rare cases the effects can be extremely serious. It is therefore safer to leave this apparently innocuous mushroom completely alone. *A. subrutilascens* has a darkish brown cap and grows under conifers in both North America and the Far East. Although some regard it as edible it is better regarded with caution as it can produce unpleasant symptoms.

The Leaden Entoloma or Livid Agaric (*Entoloma lividum* or *Rhodophyllus sinuatus*) is a large greyish-brown fungus with yellow gills which turn salmon-pink as the spores are produced. The short white stalk is smooth and curved. Although the actual poison has yet to be identified this fungus has been known to cause severe stomach pain and upset and, occasionally, permanent liver damage. It is a common cause of fungus poisoning in France particularly in southern

Above
The Yellow Stainer (*Agaricus xanthoderma*) is poisonous to susceptible people and should therefore be avoided. It has an unpleasant odour and characteristically bruises yellow on handling.

Right
This velvety white Milk Cap (*Lactarius vellereus*) of woodlands is in fact safe to eat, but another similar species of woodlands and heaths—the Woolly Milk Cap (*L. torminosus*)—*is* poisonous, and may be distinguished by its pinkish tint and shaggy or woolly cap.

and eastern parts. It grows in groups in open deciduous woodland, usually in summer and autumn but sometimes earlier in the year. Unfortunately it can be confused with edible species of white-spored *Tricholoma* but can be distinguished by the colour of its spores.

Tricholoma is a very large and cosmopolitan genus. The great majority of its species are considered safe to eat, but two clear yellow species are not. The Narcissus Blewit (*T. sulphureum*) is common, especially in oak woods, and has a strong smell described by some as of coal gas, and by others as resembling the smell of narcissus bulbs. It is best regarded as poisonous. This is certainly true for *T. aestuans*, a European species not found in Britain and rare elsewhere, growing in mossy pinewoods. It could easily be mistaken for a paler version of *T. equestre* which also grows in pinewoods but which is a valued edible species.

Mental disturbances including hallucinations are not confined to poisoning by the Fly Agaric. They can also occur after eating certain small fungi found growing among grass or on well-manured soils. Such hallucinogenic fungi include the common European species known as the Stained Gill Fungus (*Panaeolus campanulatus*). It has a greyish bell-shaped cap 3–4 cm (approximately 1–1½ in.) across with a fringed margin, and is borne on a long slender stalk up to 10 cm (4 in.) high. The flesh of the cap is reddish. This is a common fungus on animal droppings in meadows and other grassy places. If it is picked and accidentally included in a dish of mushrooms, hallucinations may develop within 15–30 minutes. Fortunately recovery usually takes place within 5–10 hours. The poison is known as psilocybin and is also present in the dainty little fungus known as Liberty Caps (*Psilocybe semilanceata*). This is common by roadsides and on rich grassland. The colour varies from yellowish-green to light brown and the caps are minute–only up to 1 cm (⅓ in.) across. They have a distinctive shape, conical and rather pointed. Psilocybin and the chemically related compound psilocin are found together in a Mexican species–*P. mexicana*. Both substances are related to muscarine, which accounts for their hallucinogenic effect.

A number of gill-bearing fungi are poisonous in varying degrees. For example the Common Ink Cap (*Coprinus atramentarius*) is generally thought to be edible like the great majority of members of this genus. If, however, it is consumed at the same time as alcohol, then well within an hour there is a marked increase in heart-beat and a feeling of intense heat spreads over the body. The effect lasts for a short time only but returns if more alcohol is taken soon afterwards. It is now known that the substance responsible is actually the same as that used to treat alcoholics by causing them to feel nausea and revulsion against alcohol. *C. atramentarius* is one of the larger species, growing up to 18 cm (about 7 in.) high, and grows in clusters usually on buried wood in fields and gardens or at the base of tree trunks. It was this species of ink cap which was used at one time to make ink.

The Woolly Milk Cap (*Lactarius torminosus*) is found especially in birch woods and is regarded as edible, but only after prolonged boiling. Its poison is still unidentified and is destroyed by heat, but when active it can cause severe gastroenteritis and may result in death. Yet after treatment by long boiling it is regularly eaten in Finland, and in Norway it is roasted and then added to coffee much as is chicory on the Continent. Probably for this reason some authorities regard it as virtually harmless. The Woolly Milk Cap has a yellowish-pink cap about 10 cm (4 in.) across with darker concentric circles and a shaggy or felted upper surface when young. Like all species of *Lactarius* it exudes a milk when broken. This is white and shows no colour change when exposed to air. *L. biennis* is common in woodlands, especially of beech, and in Britain is regarded as inedible and unsafe, yet in both France and Czechoslovakia it is regularly eaten, although only after pre-treatment in boiling water. At least five other common species of *Lactarius* are regarded in the same light.

The brightly coloured russulas sometimes look as though they could be poisonous and yet this large genus (of over one hundred species) has very few suspicious species, although some are bitter-tasting. The most well-known of these is the Sickener (*Russula emetica*). This has a bright scarlet cap which peels easily, exposing a pinkish-red layer. Both gills and spores are white as is the short stalk. It grows in all kinds of woodland between July and December. If by chance eaten raw it lives up to its name, causing violent stomach pains and upset, but has no effect if cooked. Since the acrid or poisonous substances which are present in some species of both *Russula* and *Lactarius* are destroyed by heat, all species are collected, cooked and eaten in some parts of the Continent, more especially in Poland and the U.S.S.R.

Several species of Parasol Mushroom (*Lepiota*) should be regarded with suspicion. They include the Crested Agaric (*L. cristata*) which is said to produce effects similar to mild Death Cap poisoning. It is fairly common in grassy places, including gardens, between August and November. It has a cream cap covered with reddish-brown scales and a strong unpleasant smell. *L. fuscovinacea* is also unsafe and looks unappetizing. It is found in woodlands in autumn and has a densely felted dark pinkish-purple cap and stalk. The so-called Poisonous Lepiota (*L. helveola*) is small—about 2·5 cm (1 in.) high—and smells pleasant. It grows in pastures between May and October. It has a pinkish scaly cap which reddens on bruising. Despite its rarity and insignificant size it is known to have caused cases of poisoning with symptoms of gastroenteritis accompanied by delusions.

Left
Often found in gardens, by roadsides or woodland paths the Common Ink Cap (*Coprinus atramentarius*) is noteworthy for its former use in making ink. It is edible and often picked, but alcohol should never be taken at the same time since sickness will result.

Above
The nauseating smell of the Stinking Russula (*Russula foetens*) should deter would-be pickers. If inadvertently eaten it can induce sickness.

Right
At one time thought to be virulently poisonous, the Devil's Boletus (*Boletus satanas*) should be avoided since it can cause severe gastric upset in susceptible persons.

Left
The Lorchel (*Gyromitra esculenta*) is a curious-looking Ascomycete with a squat stalk bearing a reddish-brown mop-like fertile head. It appears in spring in sandy pine woods but is never common.

Below
Although unlikely to be picked for eating, *Galerina hypnorum* is one of a number of small brown species recently found to be highly poisonous. It grows among moss in woods, heaths and bogs.

The large genus *Boletus* (with characteristic pores on the underside of the cap) is generally regarded as harmless if not always particularly tasty. One species is especially highly valued on the Continent, and this is *Boletus edulis* which the French call Cèpe and the Germans Steinpilz. Its flesh remains white even when cut, unlike the poisonous Devil's Boletus (*B. satanas*) which turns blue and pink when cut. This can cause violent enteritis, especially if eaten raw. It is found on chalky soils in deciduous woodland between July and October, but is never common. It has quite a large cap, greyish olive-green above and at first yellowish-green beneath, but becoming blood-red when mature. It is regularly collected and eaten in parts of Italy and in Czechoslovakia, but only after being well-cooked. Once again the effect of heat is to destroy the poison, although it may not eliminate the bitter flavour found in some species, for instance *B. calopus* and the Bitter Boletus (*B. felleus*).

Some of the most dangerous of poisonous fungi have received scant attention chiefly because of their insignificant size and appearance. There are species of *Galerina* which are now known to be as deadly as the Death Cap and the Destroying Angel because they too contain amatoxins. *Galerina* is a small genus of dainty brownish-yellow fungi usually growing among mosses in damp places, the best-known being *G. hypnorum*, common in woods, heaths and bogs. The Handsome Clavaria (*Ramaria* or *Clavaria formosa*) is a woodland variety, especially among beech, but is never common. It is golden or pinkish in colour, markedly branched and tufted, and grows 20–30 cm (8–12 in.) high. Only the tips of the branches are fertile. They

are highly acrid and some experts think that the fungus is edible if cooked after the tips have been removed. Even after cooking the fungus may be strongly purgative and is best avoided.

Poisonous fungi are not confined to the Basidiomycetes although they are more common in that group. The uncommon Lorchel (*Gyromitra esculenta*) is an Ascomycete and is safe to eat only after washing and then boiling. Even then it can sometimes cause severe, occasionally fatal, results. It grows on sandy soils among conifers and appears in the spring. It has a thick white convoluted stalk about 5 cm (2 in.) high and this bears a dark brown highly folded and contorted cap. As well as inducing gastroenteritis it also attacks the liver and may cause cramp, giddiness and coma. The active constituent is a relatively simple organic compound formerly called helvellic acid which is normally destroyed by the stomach acids as well as by heat. Curiously, some people may eat the fungus regularly for years before being suddenly poisoned. Other individuals are so sensitive that even handling it can cause irritation of the skin and eyes.

The False Morel (*Helvella lacunosa*), like *Morchella*, often appears on burnt soil, but is an autumn fungus. Although generally regarded as edible if first boiled, it can cause severe indigestion in susceptible persons. The dark, convoluted saddle-shaped cap gives rise to ascospores.

Above
Like all ramarias, the Fairy Club (*Ramaria stricta*) has a densely 'twiggy' growth pattern. Although not worth eating, it is harmless–unlike the coarser related species *R. formosa* which is poisonous and bitter.

Right
Ergot (*Claviceps purpurea*). The dark sclerotia of this dangerous fungus are sometimes found in the ears of cereals or of grasses, as seen here in an inflorescence of Purple Moor Grass. After falling to the ground, small fruitbodies develop from the sclerotia in the following spring.

Probably the most widespread of poisonous fungi is one which is not particularly conspicuous. This is Ergot (*Claviceps purpurea*), another Ascomycete and one which has been widely reported on throughout history. It is almost worldwide in distribution and is a parasite attacking rye and other cereals, and also grasses. Ascospores germinate on the small flowers of the host and the mycelium penetrates the ovary forming a compact club-shaped sclerotium which replaces the seed. This is purple or bluish-black in colour, about 2 cm ($\frac{3}{4}$ in.) long and is a resting or overwintering stage. If infected ears of rye are harvested the sclerotia are milled with the grain, producing an unpleasant tasting flour. At various times in history, and particularly in periods of famine or poverty, this flour has been made into bread and eaten with devastating results. The sclerotia contain a number of nitrogenous compounds such as ergometrine, ergotoxine, and ergotamine. Ergometrine has a marked effect on smooth, i.e. involuntary, muscle and causes rapid contraction of the uterus. For this reason extracts from the sclerotia were formerly used to hasten childbirth and the drug is still used to control postnatal haemorrhages. The other two alkaloids have a similar action but the effect is slower and more prolonged.

Cattle or sheep feeding on infected grain or pasture may develop scouring and convulsions, and also gangrene of the limbs. Animals which are in-calf or in-lamb lose their young prematurely, so there may be a double loss to the farmer.

These symptoms signify the disease known as ergotism which is also known in man. It exists in two forms which are rarely, if ever, found together. There is a form usually associated with lack of vitamin A in the diet which affects the nerves and brain, causing drowsiness, convulsions and hallucinations. There is also an extremely painful gangrenous type (formerly called Holy Fire or St Anthony's Fire) which is usually found in areas to the West of the Rhine. Because rye is rarely grown in Mediterranean countries, ergotism has always been restricted to central and northern Europe as well as northern states of the U.S.A. Ergot was certainly present in Iron Age crops and grasses, as has been shown by examination of the stomach contents of early man by the Danish palaeobotanist Helbaek. He has found evidence of both Ergot and barley smut. A Babylonian tablet dating from c. 2500 BC mentions 'women who gather noxious grasses', and references to ergotism may be found in the works of early Greek and Roman writers such as Hippocrates and Pliny. Several instances of what was probably ergotism are recorded in the sixth and eighth centuries, and it is interesting that the first clear account of Ergot and its properties was given by a Perso-Arabic physician as long ago as 950.

Recently it has been suggested that the famous Salem witchcraft trials which took place in Massachusetts, U.S.A., might have been the result of an epidemic of Ergot poisoning. Most of those accused of witchcraft were young girls and these are known to be more susceptible to ergotism than men. Ergot is chemically related to the drug LSD which produces similar hallucinatory and other effects. Outbreaks of ergotism associated with accusations of witchcraft also occurred in medieval Lorraine and seventeenth century Saxony. Today, with modern agricultural and milling methods, ergotism has almost disappeared. Nevertheless occasional outbreaks still occur during a succession of wet summers.

Fungi in fact and fiction

It is not surprising that plants as interesting as fungi should be the subject of conjecture and folklore. Legends about them have been handed down through the centuries. Their intriguing habit of appearing virtually overnight, and the way they seem to grow without visible means of sustenance have contributed to the aura of mystery with which they were, and to some extent still are, surrounded. They were, for example, thought to be formed by thunder–especially the truffles, about which there has always been considerable speculation. Even today the influence of thunder on the growth of fungi is still believed in some parts of the Philippines.

Possibly because fungi were, and still are so often collected for food in many parts of northern and central Europe, legends from these parts are common. One from Bohemia concerns Christ and St Peter walking through a village and begging for bread. This was given to them–both brown and white–and when they eventually reached a forest they began to eat. As they did so crumbs fell to the ground and from these grew fungi–poisonous ones from the brown crumbs and edible ones from the white.

From Yugoslavia comes a tale concerning the origin of the Fly Agaric. Whilst the god Wotan was riding his horse one Christmas night he was suddenly pursued by devils. As his horse galloped, red-flecked foam fell from its mouth and wherever the foam fell, in the following year appeared the familiar white-spotted red caps of the Fly Agaric.

In Poland the term for an old woman is a 'morchel' (i.e. a morel). This stems from the old story about the origin of the morel. One day the devil met a wrinkled old woman in a wood, and being in a bad temper killed her by cutting her into little pieces. These he scattered in the wood and each piece gave rise to a characteristically wrinkled morel.

The Ear Fungus (*Auricularia auricula*) also has an interesting legend attached to it. Traditionally, after Christ's death, Judas Iscariot hanged himself from an elder tree. In Europe this species of *Auricularia* is virtually confined to elder trees and has an ear-like shape. So it is not surprising to find the fungus is often called by the name of Judas' Ear, later contracted to Jew's Ear. The full Latin name is *Auricularia auricula-judae*.

As far back as Old Testament days the diseases of cereal crops that we know as rusts and smuts were well known and feared. The Romans believed in rust gods and they offered them propitiatory prayers and sacrifices, particularly during their grain festival in April. Primitive peoples of Mexico and Central America carved stone effigies known as 'mushroom stones'. In shape these combined a mushroom cap with a stalk fashioned to resemble a man or god. We do not know how these mushroom stones were used. Some authorities think they may simply have been utilized for grinding mushrooms, but others consider they were important in cults or religious ceremonies. Most known examples of mushroom stones appear to have been produced between 200 BC and 300 AD but in

This familiar picture-book toadstool of the Fly Agaric (*Amanita muscaria*) clearly shows remnants of the enveloping membrane as white patches on the red cap. The basal sheath is hidden in the grass. Because of its colouring it is probably the most often portrayed of all fungi, but is also well known as a dangerously poisonous fungus. It was formerly used as a stimulant.

Below
The conical pointed caps of Libery Caps (*Psilocybe semilanceata*) resemble the French Cap of Liberty. They are well known for their hallucinogenic properties.

Right
Like a number of other species, the plum-coloured *Russula atropurpurea* is common in woodlands.

Guatemala the stones are known to be very much older and were probably carved 2,000 to 1,000 years BC.

The use of fungi as stimulants, narcotics or as intoxicants in religious ceremonies is known from widely separated parts of the world. At one time several tribes in Siberia used to prepare an extract from the dried cap of the Fly Agaric. The effect of this was to induce great excitability and animation in which the affected person sang and danced, gesticulated wildly and spoke to invisible people. This was followed by a period of deep depression, and these alternating moods could be sustained by further doses of the extract. Although a stimulant, frequent and repeated use of Fly Agaric severely damages the nervous system, and its sale in the U.S.S.R. was eventually forbidden by law.

The Dyaks of Borneo and the inhabitants of New Guinea used several species of *Boletus* and also a species of *Russula* to induce a 'mushroom madness'. The practice was still current in the 1960s. In Mexico, the native Indians had for centuries used a 'sacred' fungus as a narcotic. This was reported by the Spaniards following their conquest of Mexico in 1522. The fungus concerned was probably *Panaeolus sphinctrinus* which is also found in parts of Europe. The Aztecs are also thought to have made use of two other species of *Panaeolus*. We now know that several other hallucinogenic fungi have been used in Mexico and Central America and these include the Fly Agaric, four species of *Boletus*, two species of *Russula* and as many as fifteen species of *Psilocybe*. All were taken to produce a semiconscious state and mild delirium, in which the participants sang, danced and saw visions. It is now known that these

effects were due to the hallucinogenic drugs psilocybin and psilocin and even today in a few remote villages *Panaeolus* is still taken as a narcotic. As a matter of interest one of the modern treatments of the disabling mental disorder schizophrenia is the use of the drug psilocybin.

Fungi were also made use of by primitive peoples in fertility rites. Species related to the stinkhorns (*Phallus* and *Mutinus*), but which grow in the tropics, were ground up with ashes and the resultant mixture smeared on the body during the ceremonies.

Early drawings of fungi are scarce. A painting of a 'toadstool' has recently been found in an Egyptian tomb dating from 1450 BC and there is also one in a fresco at Herculaneum which was buried with Pompeii after the eruption of Vesuvius in 79 AD. This probably depicts *Lactarius deliciosus*. The very detailed manuscripts of Dioscorides list the healing virtues of five hundred plants including a bracket fungus now known as *Polyporus officinalis*. Unfortunately his earliest manuscripts are lost but were probably not illustrated.

There is an unusual fresco in a disused church in central France. It dates from 1291 and shows a highly imaginative branched Fly Agaric which is meant to represent the tree of good and evil. It is considered by some that this fresco may illustrate that there was a link between some sort of mushroom cult and early Christianity in parts of France. The Fly Agaric is a handsome fungus and is probably the most pictured of all fungi. It is still used to illustrate fairy stories, children's books and Christmas cards.

Mushrooms sometimes figured in peppercorn rents; in the fourteenth century, the Earl of Sussex and Surrey paid for his Norfolk manor with just one mushroom a year. Mushrooms are also depicted in some coats-of-arms.

The part-medicinal part-botanical books of the sixteenth and seventeenth centuries which were known as herbals had numerous illustrations, including some fungi, although the drawings were often over-imaginative. The first known illustration of a microscopic fungus is in a book by the English scientist Robert Hooke which was published in 1665. It shows *Phragmidium*, the rust of roses. Later he described a variety of moulds on cheese, decaying leaves, roots and

wood. Because he could find no seeds, Hooke considered that fungi arose spontaneously from the material on which they grew. Later, Malpighi, an extremely observant Italian anatomist sketched a number of common moulds found on bread, cheese and fruit, and showed the spores.

In the nineteenth century the famous Victorian children's author Beatrix Potter made hundreds of exquisite drawings of fungi of all kinds. She was an excellent amateur mycologist and actually submitted scientific papers to the world-famous Linnaean Society. Her drawings remained unpublished in her lifetime, but were used to illustrate a book on fungi published to honour her centenary in 1966.

We have seen that Greek and Roman writers living before the birth of Christ wrote of the mushrooms and other fungi which they recognized. Naturally they speculated about the nature of the plants they described, and even questioned some of the beliefs such as thunder being responsible for the formation of fungi. They were puzzled by the fact that no seeds were produced and yet the fungi continued to reproduce themselves—but of course they could have had no idea of spores and never observed them. Pliny was puzzled by truffles and the way they 'can spring up and live without a root'. He believed they were formed from the soil in which they grow and thought much the same about the esteemed 'boletus' (*Amanita caesarea*), stating it arises 'from the mud and acrid juice of moist earth or of acorn-bearing trees'. Centuries later similar beliefs were still held.

In Europe in the sixteenth century and later, scientists and thinkers had more profound thoughts about the fungi. They realized that they form a large group of related organisms, and felt they might comprise a new natural kingdom between those of plants and animals. Linnaeus had much the same ideas and it is interesting that some modern mycologists hold similar views.

Fact and legend are interwoven about the well-known fairy rings. These are usually obvious circles, most conspicuous in open grassland such as pastures, downland or garden lawns, but also occurring in woodland. They range in size from a few metres to hundreds of metres across. The rings appear as a series of concentric zones, usually three zones to each ring. There is a narrow outer zone of dark green grass, an inner wider zone in which both the grass and any flowering plants present grow more luxuriantly and, between the two, a

Top
A springtime fungus of woods and hedgerows, the Morel (*Morchella esculenta*) produces ascospores in the folds of the greyish-brown head. All species are good to eat.

Above
The photograph shows part of a ring of the Fairy Ring Mushroom (*Marasmius oreades*) with numerous fruitbodies developing just outside a sparse brownish zone.

Right
The traditional fairy ring is often due to the activities of the rather small Fairy Ring Mushroom (*Marasmius oreades*) which is common in grassy places, including garden lawns.

middle zone which is more or less bare of vegetation. It is this bare zone which has given rise to so much folklore. The name 'fairy ring' is widespread, but is not universal. It is used in France, Denmark, Sweden, the U.S.A. and the Philippines. The French often use the term 'rond de sorcieres' (sorcerers' ring) and the Germans 'Hexenring'. In both Germany and Austria the rings have been attributed to witches or dragons. Other legends had it that they marked the position of buried treasure guarded by fairies or witches. Later explanations became more rational but were still incorrect. The rings were thought to be due to the effect of lightning or whirlwinds, or of activity by ants, moles or starlings, or more simply the result of tethering grazing animals such as goats or horses, and the subsequent manuring of the soil. Only towards the close of the eighteenth century was the true explanation put forward by the British botanist William Withering.

He stated that the rings were caused by the growth of the Fairy Ring Mushroom (*Marasmius oreades*) and he measured rings of various sizes, the largest being 5·5 m (18 ft) across. He also noticed that in the bare zone of the ring he found mushroom spawn but not in either of the other zones. Other botanists followed up this early work by observing that at certain times of the year groups of fungi appeared, and these were always associated with the outer dark zone. Rings were found to be caused by other fungi, including Field Mushrooms, Giant Puff Balls and species of what are now known as *Lepiota* and *Tricholoma*, and the rate of growth of some of these rings was measured.

Today we know that at least one hundred species of higher fungi form rings, and indeed one authority asserts that possibly most of the larger saprophytic fungi form them. They begin, as do all fungal mycelia, with the germination of a spore. This produces the hyphae or threads of the mycelium which radiate outwards from the central point—the spore. At first the mycelium makes no perceptible impact on the grass but gradually changes occur which affect the overlying vegetation. The mycelium draws on nourishment present in the soil and as it does so it subtly changes the nature of the soil constituents, acting particularly on the organic matter. Carbohydrates are broken down and completely absorbed by the mycelium, whereas proteins are partly absorbed and partly broken down in stages to ammonia. Soil bacteria then act on this, building it up either into ammonium salts or by stages into nitrates. The outer dark green zone is therefore a region of intense activity in the soil because of the combined effects of the growing mushroom mycelium and the soil bacteria.

Above
Russula is a large and cosmopolitan genus with a number of attractively coloured species such as *R. parazurea*.

Left
The small Dog Stinkhorn (*Mutinus caninus*) grows from a smaller 'egg' and has an odour far less strong than that of the Common Stinkhorn. Related species have been used in the tropics for fertility rites.

Right
A picture of the well-known Saffron Milk Cap (*Lactarius deliciosus*) is in a fresco at Herculaneum in southern Italy, which was buried in an eruption of Mount Vesuvius in 79 AD.

A somewhat similar explanation can be made for the inner dark green zone. Here there is a mass of old mycelium, which may be as much as 30 cm (1 ft) deep, together with the remains of old fruitbodies and dead grass. All these plant remains are acted on by the soil bacteria, as happens in the outer zone, so that once again there is a marked enrichment of the soil, particularly with nitrogen salts, and the vegetation growing in it becomes obviously more vigorous. But what of the intermediate zone? Typically this is bare, or has only sparse vegetation; however, some common species such as the Parasol and Field Mushrooms do not produce this bare effect. In this zone the hyphae have grown into an interwoven mass which has depleted the soil and filled the air-spaces between the soil particles. Thus the mycelium has the unfortunate effect of both absorbing what available water is present in the soil whilst at the same time reducing the soil's porosity so that rainwater can no longer be absorbed. The grasses and other plants growing on the surface are deprived of water and either become stunted or die, and so the familiar fairy ring is formed.

Weather conditions, especially warmth and moisture, will influence the production of fruitbodies within the ring, which may produce successive crops or may be barren. For example, a ring of *Clitocybe gigantea* was seen to produce only one crop in fifty years. In Holland the activity of a ring of Wood Blewits (*Tricoloma nudum*) was followed over three years. When first observed it measured approximately 2·1 m (7 ft) across and produced 24 fruitbodies. Twelve months later it had increased to 3·75 m (about 12 ft) across with 153 fruitbodies and by the end of the third season was nearly 5·35 m ($17\frac{1}{2}$ ft) across with a crop of 232 specimens. By contrast, observations on a ring of *Marasmius* showed an annual increase in diameter varying between 25 and 65 cm ($8\frac{1}{2}$ in. to 22 in.). This difference in the rate of growth is almost certainly due to variations in weather, particularly the amount of rainfall. In a dry season little growth takes place. Measuring the growth of rings over a period of years enables their age to be calculated. In Colorado, U.S.A., certain rings were calculated to be over 400 years old and fragments possibly about 650 years. In Britain a circle of *Clitocybe geotropa* is known to be about 700 years old. On the South Downs there are rings of the St George's Mushroom which are extremely large and which have existed for several centuries. Aerial photographs of open spaces such as downland show up the rings particularly well, but rings in woodland areas can be seen only when the fruitbodies develop.

One of the curious features associated with certain fungi is the luminescence developed either by the fungus or by decaying wood on which it may be growing. This phenomenon has been described by classical writers such as Aristotle and Pliny and so has obviously been known for many hundreds of years. It is best shown by the Honey or Bootlace Fungus (*Armillaria mellea*), which is probably responsible for most instances of luminous wood in the north temperate zone. The Honey Fungus is abundant on tree stumps and is also the cause of a damaging white rot of timber. It is common in both temperate and tropical areas throughout the world and spreads through both living and dead wood, passing from one source to another by means of cord-like strands—the rhizomorphs. These, when young, together with the mycelium, are luminous.

Other fungi with this special property of luminescence include species of *Pleurotus*, of which the best known in Europe is *P. olearius*. Britain and North America both have a gill-bearing fungus—*Panus stypticus*—which is common on old

tree stumps and which forms small clusters of cinnamon funnel-shaped fruitbodies. This exists as two different strains—the American one which is luminous and the British one which is not. In this fungus it is the mycelium, the cap and especially the gills, which are luminous. Also found in America is the well-known fungus Jack-o'-Lantern (*Clitocybe illudens*) which may be found growing in clusters on tree stumps, especially oak. It is a common woodland species with a golden-orange cap and paler gills, and is luminous when the spores are ripe.

Forestry and timber yard workers have long been familiar with the phenomenon of luminous wood since the glow from infected timber is very obvious in darkness. Incidents relating to it are known from both world wars. In World War I soldiers at the front used to fix pieces of rotten wood on their helmets or rifles to prevent collisions in the dark. In World War II it is reported that wood in a timber yard near London had to be covered during air-raids because it glowed so brightly on dark nights. Luminescence in mines is also known, and is due to pitprops and

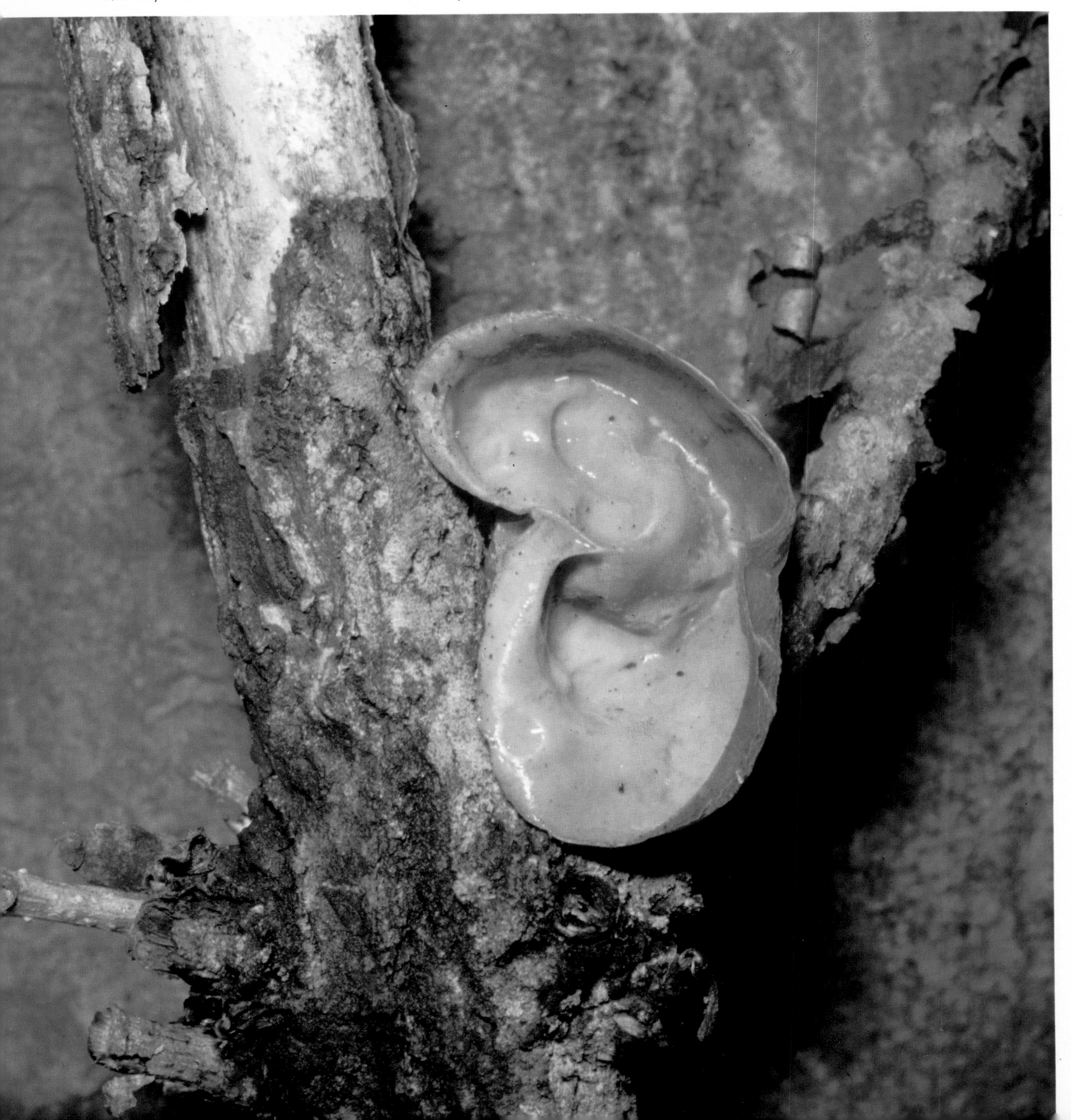

Ear Fungus (*Auricularia auricula*). The specimen shown here really does live up to its common name. This is one of the jelly-fungi although when dry it becomes hard, horny and dark in colour.

Left
Luminous fungi are common in the tropics, and they emit a surprising amount of light. This photograph was taken at night in Malaya, using just the light emitted by the toadstool.

Below
Here can be seen the typical appearance of grass infected with Stem Rust (*Puccinia graminis*). The leaves are covered with small pustules which produce orange summer spores and sap the vitality of the plant. Cereals are also attacked.

shuttering being similarly infected. Decaying leaves are sometimes luminous, and this can be attributed to the presence of a number of species of the small fungus *Mycena*. In almost all the species it is the mycelium which is luminous, and the effect is most pronounced usually shortly after the mycelium begins to grow. However, in a variety which grows in Malaya, it is the moist spores themselves which luminesce. Although luminous fungi are found in many parts of the world they are more common in the tropics and in a number of islands of the south-west Pacific. Many are species of *Pleurotus*, and they may emit sufficient light at night to see to read or to follow a path. Often they look like green eyes shining in the darkness.

Within the hyphae of the fungus is a substance known as luciferin which is acted on by a special enzyme. The chemical change which follows generates light, but the biological advantages of this to the fungus are doubtful. Luminous fruitbodies may possibly attract night-flying insects which then aid the dispersal of spores, but this does not explain the luminescence of *Armillaria* rhizomorphs.

Some fungi form close associations with certain groups of insects and it is clear that these associations must have evolved over a very long period of time since both of these groups arose very early in the earth's history. The most bizarre

of these examples of symbiotic partnerships is that shown by the formation of fungus gardens by termites and ants. The termites concerned are found in the tropical zones of Asia and Africa and are notorious for the immense amount of damage they cause. Characteristically they build enormous nests of hardened earth, roughly cylindrical in shape and nearly 10 m (33 ft) high. These nests are honeycombed with tunnels, together with specially built cavities which house the fungus gardens. The cavities vary in size from a few centimetres to half a metre in length and contain layers of droppings deposited daily by the worker termites. On this the fungal mycelium grows and produces numerous small white masses of fungus cells on which the worker termites feed. Periodically the termites clear out masses of mycelium, probably to keep the tunnels clear, and this may produce small fruitbodies of the cap-and-stalk type. But the most intriguing phenomenon is the occasional production of a truly gigantic fruitbody which grows up through the earth of the nest. This may have a cap as much as 60 cm (nearly 2 ft) across, borne on a stalk up to 2 m (6½ ft) high and looking like a sunshade on top of the nest. Such giant fruitbodies are never found away from the nest and their production poses some interesting questions. For example, does this close association with the termites provide some stimulating growth substance?

By contrast the fungus gardens of ants are formed below ground and may be found as much as 5 m (16½ ft) deep, and spreading over an area of 100 m² (1,076 ft²) or more. The ants concerned are leafcutting species of tropical America. These are so destructive that by their combined efforts they can defoliate a large tree overnight. Parts of flowers and pieces of leaf up to 2 cm (about ½ in.) across are cut and carried by the worker ants into the fungus chambers. Here the leaf fragments are chewed and mixed with ant droppings. On this steady build-up of organic matter the fungus mycelium thrives and is actively tended by the ants. They weed out unwanted plants and also transplant pieces of mycelium to new fungus chambers. As they do so their saliva falls on the fungus gardens and this has the effect of promoting the growth of the wanted fungus and yet inhibiting any invasion by other fungi. The ants feed on the fungus, nibbling the swellings which are produced at the tips of the hyphae. When young queen ants leave the nest to form new nests of their own, they carry with them in a special pouch small portions of fungus, and thus the association between ant and fungus is carried on.

Yet another of the curiosities of the fungal world is the small group of some fifty species known as carnivorous or predacious fungi. These are all simple moulds living in soil or in water. They feed on small animals such as eelworms (nematodes) which they actively catch by a variety of mechanisms. The simplest of these is a kind of sticky fly-paper. The fungal threads secrete a substance which firmly holds any animal that comes into contact with it. Following this, small absorbing filaments are injected into the animal's body which is then slowly digested. This kind of trap works well for sluggish unicellular animals but can also catch some of the more vigorous small animals which live in the soil.

Among the water moulds is one which has a kind of fishing line. Along the length of its filaments are small sticky pegs. Should one of the group of small animals known as rotifers bite one of these pegs, it becomes 'hooked' and its body is then slowly digested by the fungus. Instead of sticky filaments some fungi have sticky spores. These

Left
An irregular but complete ring of fruitbodies of a species of *Hygrophorus*.

Right
The large species *Clitocybe geotropa* grows in woodland and grassy places and may form rings. In Britain one of these rings is known to be 700 years old.

Below
The Honey or Bootlace Fungus (*Armillaria mellea*) is a highly destructive fungus which attacks and kills all kinds of trees and shrubs, spreading underground by cord-like rhizomorphs and producing huge tufts of honey coloured caps. *Armillaria* is the chief cause of luminous wood in temperate climates, both the young mycelium and the rhizomorphs emitting light.

either adhere to, or are eaten by minute soil animals and the spore then germinates, producing new spores whilst feeding on the body of its victim. Other carnivorous fungi produce small circular side-branches or rings along their hyphae. Each of these rings has three sensitive pads which swell on contact. If an eelworm should accidentally wriggle through a ring the pads suddenly swell up, inflating the ring, and the animal is squeezed and tightly held. The eelworm eventually dies and its body contents are absorbed by digestive hyphae. Sometimes this process is shortened by the production of a poisonous substance which kills the eelworm. This substance is extremely potent and tests have shown that it is effective in dilutions of one part in five million.

The trapping devices shown by carnivorous fungi evolved much earlier in time than those of the better known carnivorous or insectivorous plants, and nor are they merely a botanical curiosity. Because of their activity they maintain some control over eelworm populations in the soil and thus help to keep the roots of cereals, potatoes and other crops relatively free of these serious soil pests. Eventually the carnivorous fungi may be developed as a method of biological control over these and other pests.

The significance of fungi to man and their beneficial and harmful activities have already been touched on in earlier chapters. At times their impact on man has been epoch-making, completely changing the course of history. Thus wood-rotting fungi were the scourge of the British navy from the time of Henry VIII to the building of the first ironclad ships in the 1860s. Between wars the wooden ships were moored and slowly rotted away. During the American War of Independence sixty-six British ships sank because of decaying timbers, the best known being the 100-gun vessel the *Royal George*. This sank in Portsmouth harbour in 1782 with the loss of Admiral Kempenfelt and several hundred of the crew.

Less well known is the collapse of a projected invasion of Turkey in 1722 by Peter the Great of Russia, whose cavalry were totally incapacitated by an epidemic of ergotism. Reference has already been made to the possible influence of ergotism on the practice of witchcraft in the Middle Ages in many parts of Europe and later in America. But ergotism may have been responsible for other tragedies. For example it has been suggested that the mystery of the sailing ship *Marie Celeste* might be explained as a mass infection by ergot-contaminated bread. The afflicted crew could have suffered some kind of mass hysteria which drove them all overboard, leaving a deserted ship to sail on alone.

The most dramatic of all fungal epidemics was the Potato Blight (*Phytophthora infestans*). This was first seen on a large scale in Belgium in August 1845 and it quickly spread to all parts of northern Europe. By September of the same year it appeared in Ireland and spread with alarming rapidity. Ireland had always been a poor country, lacking natural resources such as iron and coal, and was heavily dependent on an agricultural economy based on grass and potatoes. The latter formed the staple diet of the Irish and when, in the following year, virtually the whole potato crop failed the effect was devastating. Potato Blight thrives in moist summers and is much less serious when the growing season is dry. The 1840s were characterized by a succession of wet summers suitable for the rapid spread of the disease, and between 1845 and 1851 nearly one million people died of starvation in Ireland. Although in England the Prime Minister Robert Peel used the potato famine as an excuse to repeal the Corn Laws, thus allowing the import of cheap grain, the handling of the whole tragedy was markedly inept. One and a half million Irish emigrated to the U.S.A., substantially altering the composition of the population at that time, and with repercussions that continue to the present day.

Left
The white markings on a black cap make the Ink Cap (*Coprinus picaceus*) look rather like magpie plumage—hence the specific name. All ink caps have long thin gills which produce black spores and, like the Fly Agaric, are often used by artists to illustrate fairy stories.

Above
The Yellow-stemmed Mycena (*Mycena epipterygia*) is found in temperate climates, but a number of species are tropical and all have much the same form. Many of them are luminous.

Right
Dry Rot Fungus (*Serpula (Merulius) lacrymans*). Some idea of the destruction caused by this most dreaded of fungal rots is shown here. The fungus spreads by rhizomorphs and later produces rather shapeless, white-margined brownish fruitbodies.

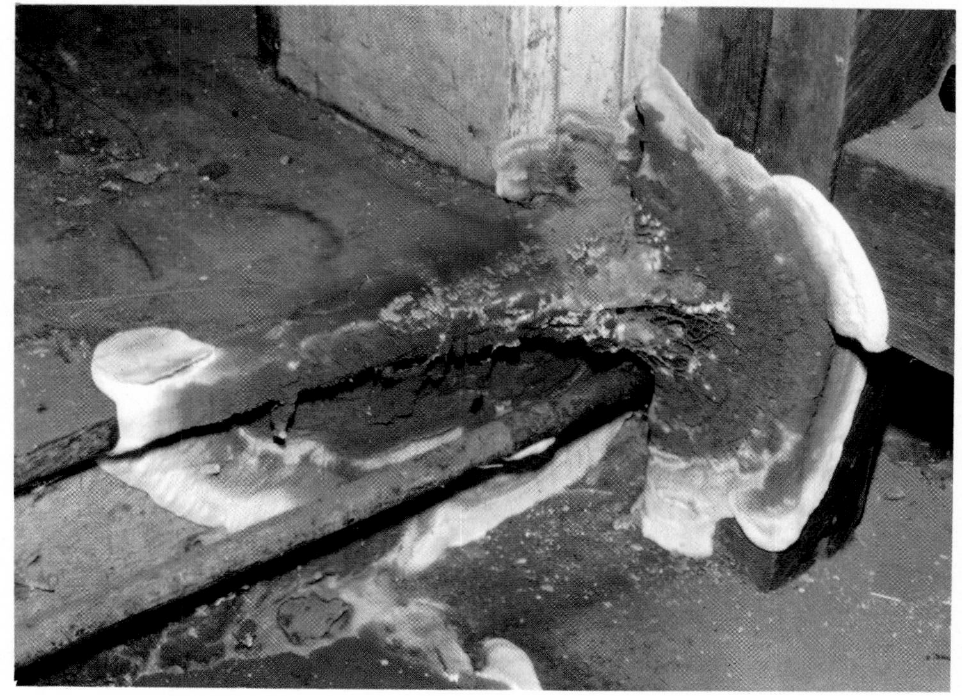

On a somewhat smaller scale the Downy Mildew of grapes (*Plasmopara viticola*) spread from America to Europe in 1876. In the New World it was a mild pest since the vines had become resistant, but introduced into another country it became virulent and by 1882 was rampant in vine-growing areas throughout France. By good fortune it was discovered that the disease could be controlled by spraying with a mixture of copper sulphate and lime and this remedy, called Bordeaux mixture after the locality where it was first used, is still an effective fungicide today.

Fortunately for modern man epidemics on this scale are unlikely to happen in modern times, but even in the twentieth century several important fungal diseases have made an impact. Thus in 1932 the

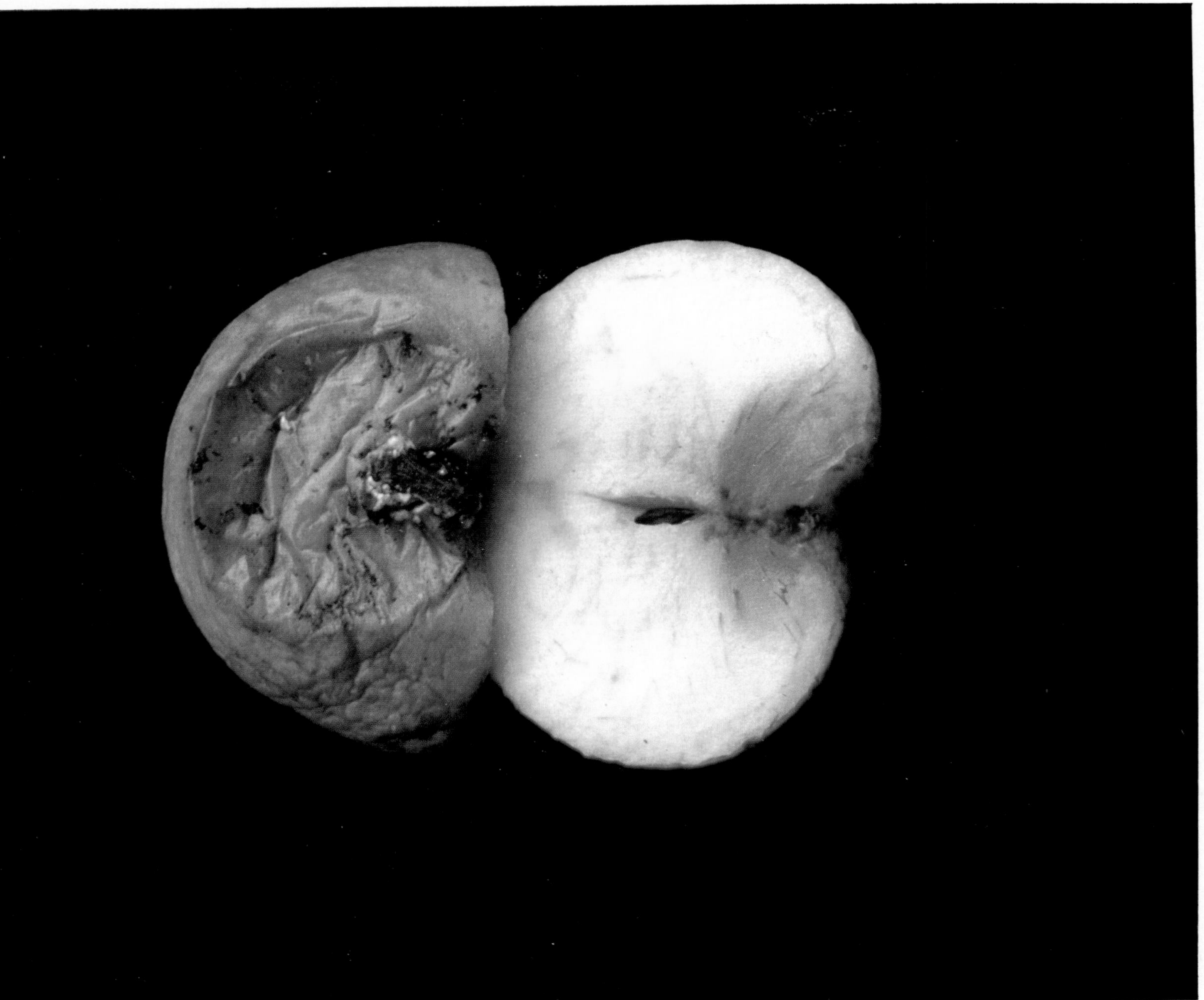

eelgrass (*Zostera*) which grew in abundance in the shallow coastal waters on either side of the Atlantic Ocean was virtually wiped out by a small parasitic fungus. The sequel was a series of very significant changes, especially in the marine life of those coastal waters, and these in turn had a direct effect on the fishing industry. A few years later the sweet chestnut (*Castanea sativa*), which formed great forests in eastern North America, was attacked by a fungus introduced from Asia and was almost wiped out. In the early 1970s a virulent strain of Dutch Elm disease (*Ceratocystis ulmi*) was introduced into Britain with a consignment of elm wood imported from Canada, and infected with a bark beetle which carried the fungus. The disease has now reached the point where the only satisfactory method of control is the felling and burning of diseased trees, and this will cause a permanent change in the English landscape as well as destroying useful sources of timber.

Fungal attacks on herbaceous and crop plants are generally reasonably amenable to control by a wide range of measures. We may also be able to use the antibiotic secretions of the fungi themselves, as well as relying on the skill of the plant-breeder to produce resistant strains of plants.

In the future man may hope to utilize the great versatility of microfungi to produce food for the world's ever-increasing population and thus harness these, perhaps the most fascinating of all plants, still more in the service of mankind.

Top left
The blackened potatoes from a plant infected with Potato Blight (*Phytophthora infestans*) are quite unfit to eat, but the disease can be checked before this happens by spraying with Bordeaux mixture.

Bottom left
The common Brown Rot (*Sclerotinia fructigena*) of apples and other fruits can cause considerable economic loss. Airborne spores may enter bruised or damaged fruit both on the tree and in store.

Below
The characteristic appearance of a blue-green mould densely powdered with chains of spores is shown here. The oily looking droplets are exuded by the fungus and contain penicillin. The commercial production of this most valuable antibotic is based on selected strains of *P. chrysogenum*.

Index

Page numbers in italic refer to illustrations